QUEST FOR SPACE

THE INDIAN CONNECT

QUEST FOR SPACE

THE INDIAN CONNECT

Brigadier **Rajat Jairath**

Introduction by

Air Marshal **Vinod Patney** SYSM PVSM AVSM VrC (Retd)

KNOWLEDGE WORLD

KW Publishers Pvt Ltd
New Delhi

in association with

Centre for Air Power Studies
New Delhi

2011 FPBA BEST PUBLISHERS AWARD (ENGLISH)

The Centre for Air Power Studies is an independent, non-profit, academic research institution established in 2002 under a registered Trust to undertake and promote policy-related research, study and discussion on the trends and developments in defence and military issues, especially air power and the aerospace arena, for civil and military purposes. Its publications seek to expand and deepen the understanding of defence, military power, air power and aerospace issues without necessarily reflecting the views of any institution or individuals except those of the authors.

Centre for Air Power Studies
P-284, Arjan Path
Subroto Park
New Delhi 110010

Tele: (91-11) 25699131

E-mail: capsnetdroff@gmail.com
website: www.capsindia.org

ISBN 978-93-83649-02-0

Published in India by Kalpana Shukla

KNOWLEDGE WORLD

KW Publishers Pvt Ltd
4676/21, First Floor, Ansari Road, Daryaganj, New Delhi 110002
✆ +91 11 23263498 / 43528107 ✉ knowledgeworld@vsnl.net • www.kwpub.com

Printed and bound in India

Contents

Acknowledgements

No words can express my heartfelt gratitude to the late Air Commodore Jasjit Singh, ex Director General, Centre for Air Power Studies (CAPS), whose very first advice to me on my interaction with him was to "add breadth at the cost of depth" to the chosen subject as not only was the subject complex but also had very little written about it in the country. His mentoring thereafter will be remembered for the free-flowing ideas and discussions that he could trigger at the drop of a hat. This work, Sir, is a tribute to you and the academic freedom that you encouraged in our esteemed Centre.

My heartfelt thanks, in equal measure go to Air Marshal (Retd) Vinod Patney who took over the mantle of Director General of CAPS, for his encouragement, guidance and insightful critique during the run up to the completion of my project. My peers at CAPS have spent many hours in 'no holds barred discussions' and peer reviews on a subject 'forced' on them, to whom I owe my deepest appreciation. It definitely made me wiser and open to critique. Many thanks to the extremely supportive staff of the library and the administration, who have played a large part in bringing my work to fruition.

I dedicate this work to my parents, Roshan and Nirmal Jairath, whose absolute faith in my ability to finish my work on my terms kept me egged on. In my wife, Ashima, I found my strongest supporter, who was remarkably patient during the endless bouts of writer's block that I experienced every now and then, allowing me to leave the home-front in her capable hands. While I searched the heavens, she kept my feet firmly planted on the ground. My at-home editor team comprised my father and my elder son, Rohan, while my younger son, Sahil bailed me out in numerous computer related crises.

This work is an ode to you all who kept me inspired in this project from its conception to its completion.

Introduction

Vinod Patney

Humankind has always had a fascination for space. Even 'moon gazing' has varied connotations. Hence, it was with great pleasure and a vicarious sense of achievement that earth-bound humans welcomed the flight of the Sputnik in October 1957. The general euphoria was also marked with some consternation in the USA. The period was at the height of the Cold War and the military advantages of occupying the 'high ground' needed no emphasis. It is no coincidence that within a few weeks, the USA fielded the very secret National Reconnaissance Organisation. The battle for space supremacy was well and truly joined.

In a few short years, Yuri Gagarin became the first man in space and the progress in development of space capabilities accelerated, culminating in Neil Armstrong becoming the first man to set foot on the Moon.

The next obvious progression was to see more from the high perch position. A number of satellites were launched and with growth in technology, reconnaissance from space was given a strong push. All-weather capabilities and increasingly better resolution soon followed. The claim that licence plates on cars could be read from space was not too far-fetched. For military planners, the capability was the fulfillment of a cherished dream.

Progress in space technology accelerated at a faster pace and underwent a series of rapid transformations. The transformations are continuing at still greater pace. Coincident with space technology, data processing capabilities also improved and a sea-change in exploitation of space was brought about

Air Marshal **Vinod Patney** SYSM PVSM AVSM VrC (Retd) is one of the most distinguished fighter pilots in the Indian Air Force's history: a decorated veteran of the 1965 and 1971 Wars. He was awarded the Sarvottam Yudh Seva Medal (SYSM) for spearheading air attacks during the Kargil War, making him the sole recipient of the SYSM till date in the IAF. Currently, he is the Director General, Centre for Air Power Studies, New Delhi.

with the ability to position Geostationary Earth Orbit (GEO) satellites. A mere three satellites could cover the globe and communication systems got a fillip. Also, physical exploration of the Moon and Mars is now a reality and efforts are on to find planets that will permit human habitation. Our space vehicles have also travelled to the far reaches of our solar system, sending back signals and photographs where possible. Recently, one of our space vehicles broke through the barriers of our solar system in an effort to explore even more distant areas. Earlier, the International Space Station (ISS) became operational, facilitating long duration stay in space, space walks and experiments of different types. The Hubble telescope was very useful to view the stars and such distant objects. There is little doubt that, with time, we will see even greater exploration and exploitation of space.

To begin with, space exploitation concentrated on military uses. Militarisation of space followed and was universally accepted as a norm. Capabilities were created for reconnaissance in its various forms, and better and faster communications permitted the use of the Internet—a capability originally intended for the US military and now a household name. The USA was the first to launch the Global Positioning System (GPS) and similar systems have been developed or are under development by other countries. GPS has been a boon for better navigation as well as for better accuracy of weapon delivery. The term 'precision strikes' was born. At the same time, GPS capability, coupled with a far more efficient Internet system permitted network-centricity and the term Network-Centric Warfare (NCW) came into being. As it always happens with military capability, the downside was the creation of systems to undermine the greater capability afforded by space assets. Thus, the terms Information Warfare (IW) and later Cyber Warfare (CW) resulted. As always, the offensive/defensive paradigm operates everywhere and continuously. There are three technologies whose progress will have a marked impact on space exploitation. These are improvements in communications, data processing and miniaturisation. In all the three areas, that are interconnected, the rate of progress is admirable.

The commercial application of space was never in doubt. The Internet, better global communications and GPS opened the field for the civilian commercial space industry. Once again, the growth was phenomenal. In the year 2012, the commercial space revenues exceeded \$200 billion. On the cards is the search and mining of minerals on our planets and the use of a cleaner low gravity environment for cost-effective industrialisation. It is always difficult to predict the future but it can be said with some surety that it will be exciting.

Space travel is no longer the preserve of military astronauts although it will be some time before civilian enterprise can match the very praiseworthy exploits of military astronauts. In 2004, a mere 47 years after the Sputnik took to the skies, Burt Rutan's Space Ship One won the \$10 million 'Ansari X Prize' for being the first private company to go past 100 km altitude—the *Karman Line*, the arbitrarily defined lower limit of space, twice in two weeks. More manned space flights with civilians onboard have followed and now bookings are being accepted for space tourism and even a one-way trip to Mars. Commercial interests often surpass military requirements!

To date, the commercial interests of companies belonging to different nations have overlapped at times and have even been competitive but the competition has not degenerated into conflict. We cannot be sure that this will be the case in the future as well. There is an oft-repeated adage that 'conflict follows commerce'. If such a situation were to occur, and the chances increase with the greater number of players wanting a foothold in space commerce, we would be entering uncharted territory with all the pitfalls of the 'uncertain'. Will it lead to rapid weaponisation of space? Will weaponisation of space occur in any case because of military imperatives? Only time will tell, but the dangers and pitfalls are much too evident and scary. The use of Anti-Satellites (ASATs) in the past is but an indication of what could transpire when commercial or military interests are threatened. Collisions have occurred in space, either planned or accidental. Space debris that resulted is only one of many problems that could get worse. There are techniques available to alleviate the problem, including removal of space debris but the moot point is: who will pay for what will be the common good—the countries that caused the problem in the first place or all those that hope to gain from a reduced danger? No answers are forthcoming. Once again, commercial interests abound.

Space is a vast expanse but there are limiting factors. For instance, there is a finite number of stations available for GEO satellites. The height is fixed at 36,000 km and the location has to be directly over the Equator. To avoid interference, a separation is essential and there is always an optimum location that may not always be available. Already, there are 435 satellites in GEO and the slots are getting crowded. More slots may be available if technology ensures that the satellites can be placed at closer intervals without causing interference. For similar reasons, the orbit of the satellites has to be controlled and there is a large number of satellites in orbit. Again, the frequencies at which the satellites may operate has to be determined to avoid interference. These tasks are carried out by the International Telecommunications Union (TU). So far, there has not been any serious problem and ITU has been allowed to dictate terms. Will this obtain when national interests conflict with those of another nation, possibly a nation that could be a prospective adversary? There are many possible imponderables.

Space is a common heritage of mankind—a global common. The Cold War ensured that there would be rivalry in space and yet it was thought prudent to have the United Nations adopt the Outer Space Treaty. This was done very early in the Space Age, in 1967, and four other treaties were adopted by 1979. Five sets of principles that have the legal status of General Assembly Resolutions soon followed. So much has altered in the last 40 years or so and yet there has been no addition or even alteration to the space laws. If the laws did not cover enough ground and possible contingencies when they were adopted, the situation has markedly worsened now. They remain inadequate and there is no effective system for conflict resolution. It must be recognised that it is never easy to get nations with differing needs on a common platform and even more difficult to institute measures to implement the laws. The requirements are disparate and the views often selfish—the common good is acceptable to a nation only if it satisfies its own requirements. Possibly in recognition of the fact that unanimity is unlikely to be obtained for a meaningful treaty, the European Union (EU) has floated the concept of an International Code of Conduct (ICoC). The effort is laudable and should be pursued but it has to be accepted that no Code of Conduct can replace an enforceable treaty duly signed and ratified.

India ranks amongst the leading 'space' nations but falls well short of what the leading countries have achieved and where we would like to be. Yet our space organisation has a number of achievements and the Indian Space Research Organisation (ISRO) ranks amongst our success stories. The first Indian satellite, *Aryabhatta*, entered orbit in April 1975 and in July 1980, we used our own launcher for a satellite launch. We started with civilian uses of space, including education, and remote sensing. We soon became one of the leading countries in the field of remote sensing, and sterling work was done to, *inter alia*, aid agriculture. On September 9, 2012, India launched its 100th mission into space. Earlier, on October 22, 2008, we launched India's first lunar probe and using a Moon impact probe, showed that water exists on the Moon. This was an important finding and won accolades internationally. India's plans to develop an Indian GPS like system and work towards better air space management are exciting developments that promise still greater achievements to follow.

Military uses of space started on a low note and were limited to a growing but limited capability for reconnaissance and communications. Till very recently, the military piggy-backed on civilian uses and had little say in the control of satellites although we did establish reconnaissance processing facilities for analysis of information obtained from our satellites and some others. We also had the use of some transponders on our INSAT satellites for communication needs. A sea-change occurred on August 30, 2013, when GSAT 7 was launched, essentially for the communication needs of our Navy. A satellite for the Air Force and Army communication requirements is in the offing.

An impetus to military application of space is long overdue. We need satellites and ground control systems under direct military control. For reconnaissance needs, better data processing facilities are required that will also permit automatic target detection and change detection capabilities. There is also a requirement for an organisation that will oversee military use of space assets, training of personnel, stipulation of procedures and priorities, and formulation and implementation of future plans. The formation of a military Aerospace Command is long overdue.

The modernisation and expansion of the Indian space sector need periodic reviews, based on future needs and an audit of past performance. Whilst financial outlays will always have an overbearing impact, some issues must find mention in the blueprint for future growth. Some of these areas are systems that speed up policy formulation and implementation with due accountability, improvements in the quality and quantity of satellite projection systems, possible international collaborative and cooperative approaches, and a study on the future requirement of spacedromes for spacecraft. Much has been achieved but much still needs to be done, and in a competitive world, we cannot afford to rest on our laurels even momentarily.

Brigadier Rajat Jairath has done commendable work in writing this book whilst he was a Senior Fellow at the Centre for Air Power Studies. He has covered the full ambit of a study on space issues, including its exploitation, advantages and dangers. He has done well to offer broad perspectives on complex issues. The book is very readable and readily understandable. Each chapter is self-contained but the continuity of the subject is maintained. Each chapter includes historical references dating back to ancient times leading on very deftly to the current state and future prospects. The book prepares the reader for the final chapter that discusses, dispassionately, the imperatives for India. The book is recommended reading for those interested to learn about space issues and also for the more seasoned professionals and practitioners.

Vinod Patney

Director General, Centre for Air Power Studies

1

Understanding Space

Dedans le coing de luna viendra rendre,
Ou fera prins and mys en terre eftrange,
Les fruitz immeurs feront a grand efclandre
Grand vitupere a l'vn grande louane.

— *Nostradamus (1503-1566)*
Century IX, Quatrain 65

Michel de Nostredame (December 1503 – July 1566), famously known as Nostradamus, was a reputed French seer who published collections of prophecies that have since become famous worldwide. Being vulnerable to opposition and castigation on religious grounds, as prevalent in Christian Europe in the 16th century, he devised a method of obscuring his meaning by using coded syntax, word games and a mixture of other languages such as Greek, Latin and Italian. These 'coded' prophecies led subsequently to a whole host of researchers, scholars and writers to draw meaning to his text, resulting in a wide range of interpretations. Due to the subjective nature of these interpretations, however, no two of them agree on exactly what he predicted, whether for the past or for the future.

Erika Cheetham (1939-1998), an English writer, linguist and medieval scholar, with a doctorate from Oxford, translated the above quoted quatrain as, *"He will be taken to the corner of Luna where he will be placed on foreign land. The unripe fruit will be the subject of great scandal. Great blame, to the others great praise"* and further interprets to say, *"If Luna is to be understood as the Moon, this quatrain is surprisingly good. The Americans* ***landed on the Moon*** *led by Neil Armstrong in 1969. The 'unripe fruit which causes great* *scandal' is the tragic explosion of the* ***Space Shuttle Challenger*** *in 1986. The great praise goes to the others, the Russians, who have satisfactorily installed their* ***space station, the Mir****, which is in continual use".*[1]

Even if the above interpretation is assumed to be an individual writer's flight of fancy or stray musings, or even if Nostradamus had not prophesised anything remotely connected to matters of the sky or space, it was still a certainty that mankind at some point in time would look heavenwards and attempt to reach for the skies. The evolution of humanity and civilisation has been witness to the step by painstaking step that mankind took towards progress. In the context of the three dimensions, being a land-based mammal, man first moved to explore and conquer land, and then built ships and launched himself to explore the high seas and distant horizons and now, within the last hundred years, mankind has "grown wings" and has taken to exploring the skies above him. What would be mankind's next logical evolutionary step? To go above and beyond the shackles of the immediate atmosphere and explore the vastness of the next dimension—space—and that's precisely what mankind has been indulging in for the past six decades.

The realisation of this dream did not depend merely on the availability of powerful rockets to propel spacecraft into space, but, more importantly, on the understanding of space. The two pioneers in the quest for space as early as the 17th century were John Keplar, a German astronomer and mathematician, who worked out the mathematical laws that govern the motion of bodies in orbit around the Earth and other planets and stars; and Isaac Newton who unravelled the mysteries of planetary motion, the laws of gravitation and motion. From this ground breaking work, mankind progressed to reach space. This chapter attempts to lead the reader, in layman's terms, through the basics of space, the fundamental laws that govern the launch and orbit of satellites, the types of orbits, and the role of international bodies in frequency spectrum management for space. The progress of the human race to reach into space and its consequent odyssey in the last six decades is being dealt with in detail in the subsequent chapter.

Where Does Space Begin?

In astronomical terms, space is the part of universe lying outside the limits of the Earth's atmosphere. In real terms, there is no clear-cut boundary between the atmosphere and space. The atmosphere gets thinner and thinner as you go up, and even when it has become too thin to sustain the thrust, lift or control of an aircraft, it is still measurably present. Once the aircraft reaches air too thin for safe control, it passes the realm of aerodynamics and enters the realm of astronautics – if it has not yet reached space, it is definitely leaving air. Thus, the usage of conventional air breathing engines is limited to altitudes that provide air. Commercial airliners typically cruise between 30,000 ft and 40,000 ft and high performance combat aircraft might reach between 50,000 ft and 60,000 ft, or even a little higher, but very few aircraft have gone much above this. A practical limit for conventional aviation would be 100,000 ft.[2] Could 100,000 ft (30 km), thus, be the defining boundary between atmosphere and space? While aviation is limited to below 100,000 ft, the atmosphere at this altitude is still too thick for a spacecraft to achieve orbit without continuous thrust.

Neither the Outer Space Treaty, nor any other international agreement, *expressly* defines outer space. There have been periodic attempts to establish a legal definition, which plainly would have implications for overflight (flying in the air space of a non-participating country), by creating a minimum altitude at which it may take place. Opposition to such a definition is based on the belief that no pressing need exists, and that the creation of a definition may inhibit exploitation of emerging technologies. Demonstration of a practical application of capability in near space, however, may force resolution of this issue, in the times to come.[3]

The Federation Aeronautique Internationale (FAI), the world governing body, with its headquarters in Lausanne, Switzerland, for air sports, aeronautics and astronautics world records, has proposed a limit called the Karman Line[4] – an altitude of 100 km as a working boundary.[5] FAI follows this altitude limit for international standard setting and record-keeping for aeronautics (aerial activity below 100 km altitude) and astronautics (aerial activity above 100 km). Within the US aerospace programme, aviators who have travelled beyond 80 km altitude are recognised as space travellers or astronauts.[6]

Characteristics of Space

Outer space is the closest natural approximation of a perfect vacuum. It has effectively no friction, allowing stars, planets and moons to move freely along their ideal orbits. However, even the deep vacuum of intergalactic space is not devoid of matter, as it contains a few hydrogen atoms per cubic metre. The sparse density of matter in outer space means that electromagnetic radiation can travel greater distances without being scattered. While space, in the region between the Karman Line and up to approximately 40,000 km altitude (where the scope of this book lies) is characterised by numerous physical, chemical and astral properties, the most obvious peculiarity is that space is a uniquely hostile environment, for both space travellers and spacecraft. This chapter will deliberate only upon those natural facets that impact upon mankind's space travel, limited, again to astronautic activities within this band of altitude.

Space Radiation: Life on Earth by and large is sustained by energy from the Sun, while in space, the Sun is both a source of energy and a potent threat. Besides the visible light spectrum from the Sun that can be perceived by the naked human eye, space is bathed in a flux of dangerous ionising radiation, both natural and man-made. The naturally existing environment is the Earth's trapped radiation belts, the Van Allen belts (discussed separately below), solar particle events and galactic cosmic rays. These are made up of subatomic particles accelerated to intense energy levels by violent activities on the Sun, as well as exploding stars throughout the Galaxy. These radiations comprise Infrared (IR), Ultraviolet (UV), X-ray and gamma radiation, and, in addition, various energetic charged particles. Ionising radiation elements resulting from human activities include space nuclear power systems (fission reactors and radioisotopes), the detonation of nuclear explosives in the upper regions of the atmosphere or in outer space (activities currently banned by an international treaty), space-based particle accelerators and radioisotopes used for calibration and scientific activity.[7] On the Earth, the atmosphere and the shielding effect of the Earth's magnetic field protect life from the vagaries of the Sun's radiation, e.g., the ozone layer in the atmosphere protects life from UV radiation. However, the same protection is not accorded in space. These rays are extremely hazardous to living cells, and space travellers exposed to them could suffer anything from near immediate acute radiation sickness to cancers in later life. Different components of space radiation have different effects on spacecraft. While solar particles impacting upon a spacecraft's airframe could affect its motion, the continuous impact of solar particles can cause physical damage, especially to delicate components like sensors and solar panels. Individually charged particles can penetrate electronic components and could corrupt computer memory, while charged particles in sufficient numbers may induce noticeable currents or significant voltages, which on discharge could cause severe damage.

The Van Allen Belts: The Van Allen belts, occurring from 320 to 32,000 km[8] altitude are the most prominent amongst the concentrations of radiation, magnetic and charged particles that surround the Earth. Their effects are most pronounced above approximately 1,000 km, and diminish again at higher altitudes. The phenomenon is further compounded by the fact that these belts are aligned with the Earth's magnetic fields—with the North and South magnetic poles, and *not* the geographical poles. This leads to concentration of Van Allen effects over the South Atlantic Ocean, leading to what is known as the South Atlantic Anomaly (SAA), where the belts come closest to the surface of the Earth.[9] This SAA is closest southeast of the Brazilian coast. These belts manifest themselves in degrading material properties and components performance, often resulting in reduced capabilities, especially in solar cells used to provide electric power for spacecraft components, or even failure of spacecraft systems and experiments. The concentration of hazardous phenomena in the belts' structure is a serious limiting factor on orbit selection, for both manned and unmanned missions.

Temperature and Pressure: Space is essentially a cold, near vacuum like environment with reduced or nil gravity and throws up unique engineering challenges. One side of the spacecraft is exposed to the unmoderated solar radiation from the Sun, while the other side is exposed to the extreme cold of space. This sets up thermal stress across its structure. In addition, the heat generated by the payload onboard the spacecraft will need to be dissipated in the vacuum of space. Lack of air makes metallic structures brittle over time and composite materials tend to lose gases from their internal structure.[10]

Weightlessness: While the mass of an object will be the same on the Earth, in space or on any other planet, the weight of that object is defined by the effect of gravity on that object. The larger and more dense a planet, the greater would be its gravitational force. This effect of gravity diminishes with altitude, that is to say, a spacecraft orbiting the Earth at an altitude of 400 km, undergoes our planet's gravitational attraction 12 percent less than that experienced on the Earth's surface. For a spacecraft to escape the planet's gravity, it would need to achieve sufficient velocity to break away from the gravity well. The escape velocity, the minimum velocity for a flying body to escape gravity, for the Earth is 11.2 km per second while that for the Moon is only 2.4 km per second.[11] Once out of the planet's gravitational pull, the spacecraft, technically speaking, has no gravity force working against it. However, a "zero-gravity" environment practically can never ever be totally achieved in a spacecraft. The minute drag exerted by an extremely thin almost vacuum-like atmosphere, venting of gases from the space vehicle, and even crew motions inside the spacecraft create nearly imperceptible forces on people and objects alike. These tiny forces are collectively called "microgravity".[12] In this environment, space travellers and their equipment are almost, but not entirely, weightless. Weightlessness comes with its own effects on humans. The most common problem experienced by humans in the initial hours of weightlessness is known as the Space Adaptation Syndrome (SAS), commonly referred to as space sickness. Symptoms of SAS include nausea and vomiting, vertigo, headaches, lethargy, and overall malaise. The most significant adverse effects of long-term weightlessness are muscle atrophy and deterioration of the skeleton, or space flight osteopenia. Other significant effects include slowing of the cardiovascular system, decreased production of red blood cells, balance disorders, and a weakening of the immune system. Lesser symptoms include loss of body mass, nasal congestion, sleep disturbance, excess flatulence, and puffiness of the face. These effects begin to reverse quickly upon return to the Earth.

Orbital Decay: Orbital decay is the process of prolonged reduction in the altitude of a satellite's orbit. This can be due to drag produced by an atmosphere due to frequent collisions between the satellite and surrounding remnants of the atmosphere, even in the near-vacuum conditions. The drag experienced by the object is larger in the case of increased solar activity, because it heats and expands the upper atmosphere. It causes the spacecraft to slow down, which leads to lowering of altitude, taking the spacecraft down to relatively denser atmosphere, which further leads to increased drag. This feeds into a vicious cycle until either the spacecraft needs to be boosted to regain orbital altitude, or the orbit decays, causing the satellite to burn on reentering the Earth's atmosphere. It flows naturally that in order to keep a satellite in orbit at low altitude for a protracted duration, it would need to be boosted regularly to overcome drag. This carries with it a fuel penalty, which will impact either the launch mass or limit the satellite's operational life. Although some drag occurs at all altitudes, above 600 km, it is essentially negligible compared to other effects.[13]

Space Debris: Space debris, also known as space junk, and space waste, is a collection of objects in orbit around the Earth that are naturally existing in space and found transiting through the atmosphere or were created by humans but no longer serve any useful purpose. The man-made debris objects in orbit, called orbital debris, consist of everything from spent rocket stages and defunct satellites to erosion, explosion and collision fragments and are being covered in adequate detail in Chapter 5. Amongst the naturally occurring phenomena is the almost continual collision between the Earth and anything orbiting it, with meteorites and mini-meteorites which originate from elsewhere in the solar system. Space science data indicate that at any one moment, a total of approximately 200 kg of meteoroid mass is within some 2,000 km of the Earth's surface. Most of this mass is found in meteoroids of about 0.01 cm diameter.[14] Shooting stars, for example, are typically no bigger than gravel grains, burning up as they enter the Earth's atmosphere. In space, however, there is no such protective cover of an atmosphere. Although the statistical risk of a substantial impact disabling a spacecraft is low, structures must be robust enough to absorb the small impacts that are likely occurrences.[15]

From Earth to Orbit

With the launch of the Sputnik 1 on October 4, 1957, the Soviet Union ushered in the era of the 'modern space age'. The Sputnik 1, a 83.5 kg spherical satellite, launched into space by a 29-metre (m) tall R-7 rocket became the first human-made object to be placed in an orbit in space around the Earth. This set the pace for the space race that had commenced post World War II, between the two superpowers. While the historical context of the evolution of the space age is being dealt with in adequate detail in Chapter 2, this section will bring out *the theory*, in layman's terms, that goes behind launching anything in space.

Most historians agree that the Chinese were the first to produce a rocket. The first reported use of rockets in warfare, was in 1232 AD[16] when the Chinese repelled Mongolian invaders with a barrage of "flaming arrows", which were essentially solid fuel arrows powered by gunpowder. The principles by which rockets operated were not understood until much later, when the thought process began about using rockets for transporting people. Science fiction tales by Jules Verne in the 19th century, proved to be a strong stimulus for each of the three great space-travel visionaries of the 20th century: Konstantin Tsiolkovsky, Robert Goddard and Hermann Oberth.[17]

Fuelled by the desire to explore space, visionaries and other members of the scientific community began to understand and harness the basic rules and principles enunciated by thinker-scientists of the past. Johannes Kepler (1571-1630), a German astronomer-mathematician, developed the three laws of planetary motion, which gave a contemporary mathematical context to modern astronomy. It was he who derived that the orbits of all planets around the Sun are elliptical and also established an empirical relationship between a planet's orbit and the distance of the planet to the Sun.

Sir Isaac Newton (1642-1727), was a British astrophysicist and mathematician, who, about a century after Kepler's initial formulation of the three laws of planetary motion, introduced the law of gravitation and codified the three laws of motion that govern the behaviour of bodies at rest and motion. The law of gravitation states that every particle of matter in the universe attracts every other particle with a force that is directly proportional to the product of their masses and inversely proportional to the square of the distance between them.

Newton's First Law of Motion states that a body in a state of rest or motion continues to stay in that state unless acted upon by an external force. But our everyday earthly experience tells us that a bullet fired from a gun, loses speed. If it doesn't hit the intended target, it will lose its momentum and eventually drop to the ground. On the Earth, the bullet loses momentum because of friction, air resistance and gravity. This does not hold true in the vacuum of space, where we approach the Newtonian ideal and motion continues unchecked.

Newton's Second Law of Motion states that the rate of change of momentum of a body is directly dependent on the force applied on it.

Newton's Third Law of Motion brings out the relation between action and reaction, wherein, for every force acting upon a body, there is a corresponding force of equal magnitude exerted by the body in the opposite direction. This law works very well to explain the functioning of a rocket. The 'action' is the combustion-of-fuel causing hot gases escaping rearwards from the rocket. The 'reaction' is the motion (of equal magnitude) of the rocket in the opposite direction.

Orbital Motion: The rules and laws given above come into play to place an object in space. To fathom the concept of an object requiring the requisite velocity to be shot into space, imagine a cannon on a mountain top. If the cannon is fired horizontally, that is, parallel to the Earth's surface, the shot would travel forward and away from the cannon. The shot will also continue feeling the downward-pulling force of gravity. While the downward gravitational pull would be constant throughout the flight of the shell, the forward motion would continue to decline. The shot would follow a curved path, a trajectory, downwards till finally landing on the surface of the Earth, defining some range achieved by the shot. The faster the shot emerges from the cannon, the more will it travel forward, thus, achieving more range, before once again hitting the ground. In both the cases, the time taken from firing of the shots to hitting the ground, surprisingly, is the same. This is so because the vertical acceleration towards the ground due to gravity, in both the cases, remains the same, a constant at 9.8 m per second (at sea level). The flight time being the same for both, it is the greater forward velocity of the latter that gives it a greater range. In a technical sense, the trajectories followed by the shots are parts of an elliptical path around the centre of gravity, the elliptical path which was interrupted by striking the Earth. To take this further, imagine colossal speed and range being imparted to the next shot. While it is in flight, the Earth would be curving away below it.

The Earth's surface is curved, such that over 8 km in any given direction, the surface falls away by about 5 m.[18]

Fig 1.1: Earth's Curvature

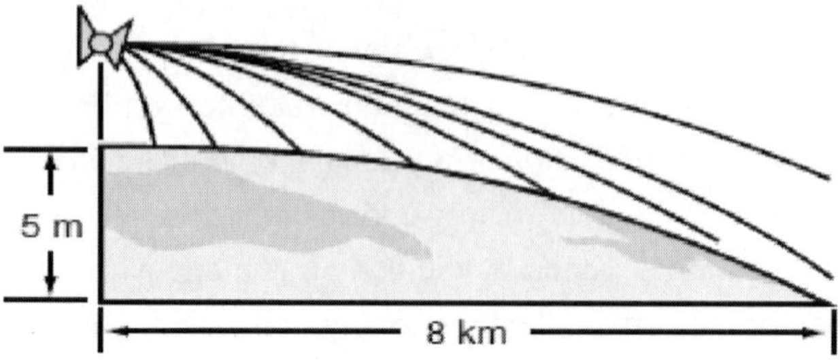

Adapted from http://space.au.af.mil/au-18-2009/au-18_chap06.pdf on December 28, 2011

Suppose that an object was moving horizontally at about 8 km per second as it began to fall. Over the 8 km distance travelled in one second in the horizontal plane, the object would drop 5 m in the vertical plane, but the Earth would curve away by 5 m because of its shape. Thus, at the end of that second, the shot would still be the same height above the ground. Assuming we can combat air resistance along both the horizontal and vertical directions, the required orbital velocity for an orbit just above the surface of the Earth translates from 8 km per second to just under 29,000 km per hour, and one orbit of the Earth would take about 84 minutes.[19] Adding the component of air resistance to this orbiting object, the height of 150 km is found to be a minimum practical one to orbit beyond the denser parts of the Earth's atmosphere. At this height, a satellite needs to travel at 7.8 km per second to attain a period of 87.5 minutes.[20] It is the job of the rocket to carry the satellite straight up through the dense part of the atmosphere and release it at its orbital altitude. While in orbit, if the satellite is speeded up, it goes into a higher orbit and if slowed down, it drops into a lower orbit.

Satellites

A satellite can be defined as any object, either man-made or naturally occurring, that orbits around something else. For example, the Moon orbits around the Earth and is, thus, a satellite. The Earth orbits around the Sun and is a satellite of the Sun. Other examples of naturally occurring satellites include comets, stars, asteroids, and other planets. Man-made satellites placed around the Earth are called artificial satellites, or more commonly satellites. While the orbiting object is the secondary body, the Earth, or any other planet, would be the primary object.

Man-made satellites are classified by type, and there are over a dozen types. Some of the main types of satellite used today are for astronomy, communications, Earth observation, weather and as space stations. Other types of satellites include those that are used to monitor the Earth from a military standpoint, and biosatellites which may carry animals or other life forms for the purpose of research on Earth life forms in space. Navigational satellites are yet another type which form a vital part of the US' Global Positioning System (GPS), an application commercially available in many cars and hand-held mobile communication sets.

Regardless of the classification of a satellite, common to all of them are factors that govern their practicality. Footprint is one, which is the area of the Earth under it, potentially 'visible' to it. The revisit time of a satellite defines how often a satellite overflies a given point on the Earth. For a practical mission, this often dictates the required orbit. The higher the altitude of orbit, the greater the period (time taken by a satellite to travel once around its orbit), and the greater the interval between two successive revisits, though the longer is the duration of each overflight too.[21]

Launch Considerations

Having had a brief look at what a satellite is and what the theory behind placing it in orbit is, this section will, in general terms, familiarise the reader with the science of launching it from the Earth, i.e., the practical dimensions of gaining access to space. Conventional aircraft engines cannot reach space due to their dependency on oxygen. The current and only available practical means of reaching space is a rocket.

Rocket: In general, a rocket could be defined as a completely self-contained projectile, pyrotechnic device, or flying vehicle propelled by a reaction (rocket) engine, obeying Newton's third law of motion of *action-reaction*. Since it carries all of its own propellant, including its oxidiser, a rocket vehicle can function in the vacuum of space. Rockets can be classified by the energy source used by the reaction engine to accelerate the ejected matter that creates the vehicle's thrust, such as nuclear, electric and chemical rockets. Nuclear or nuclear-thermal rockets derive their propulsive thrust from nuclear energy sources by using a nuclear-fission reactor to start, maintain and control a nuclear-fission chain reaction, to burn the propellant (generally hydrogen). For the same amount of total propellant mass expelled from the nozzle, the nuclear rocket gives more than twice the total thrust, though nuclear radiation hazards limit their employment to regions much beyond the Earth's biosphere.[22] The nuclear rockets are, thus, most suited for inter-planetary missions which warrant fast flight times and larger payloads, and are likely to be the rockets of the future for deep space exploration. The electric rocket employs electric energy and converts it into a forward directed force to attain the required exhaust velocity. The thrust produced is relatively lower and such rockets are ideally suited for non-priority payload delivery applications where constant and gentle thrusting over long periods is the need. Chemical rockets rely on chemical reactions to produce the desired thrust and, in turn, are divided into two general sub-classes: solid-propellant or solid-fuel rockets and liquid-propellant rockets.[23] There are technical advantages and disadvantages for each type, and modern launchers even use a combination of solid and liquid fuels. For their weight and/or volume, solid-fuelled rockets usually produce significantly more thrust than the liquid-fuelled ones. Also, as the oxidiser and the fuel are precombined in the solid-fuelled rocket, there is no need to incorporate complicated pumps and valves to control the flow of volatile liquid fuels. They are much simpler and inherently more reliable, though they suffer from one major disadvantage, in that, once ignited, they cannot be extinguished nor controlled nor shut down in flight. Thus, a system using only solid fuel for its rocket, will have some amount of uncertainty in its final position or speed in orbit. Liquid-fuelled rockets, on the other hand, are more complex but offer the tremendous advantage of allowing control of the thrust, by regulating the flow of fuel and oxidiser, enabling them to be shut down and reignited in flight.[24] Rockets in launch vehicles could be either expendable, or reusable as in the case of the Space Shuttle. The Space Shuttle, the commonly used generic name for the National Aeronautics and

Space Administration (NASA's) Space Transportation System (STS), is unique in that it is a delta-winged aerospace vehicle that launches into space like a rocket and for its return, uses its aerodynamic delta wings to glide through the atmosphere and land on a runway.

Multi-Stage Launch: A simple rocket suffers from the disadvantage of being a dead weight once its fuel is expended. It, thus, needs to be jettisoned, to enable only the payload to be taken to its destination. Thus, the concept of multi-stage launcher gains ascendency. Multi-rockets, usually of reducing size, are stacked one on top of another; the first stage is used at the initial launch to accelerate the whole system clear off the ground. When its fuel is consumed, the first stage separates and the next stage ignites and accelerates the much smaller (less massive) system on to the next stage of its journey. Since total mass is much reduced, greater acceleration is easier to achieve. Use of up to three stages is common;[25] where necessary, the first stage can be augmented by solid-fuel rocket boosters. Using them in conjunction with the first stage gives maximum scope for any correction required due to their imprecise thrust output, and uses their high thrust level when the mass is greatest. Calculations show that for a given total mass at launch, a two-stage launcher could typically deliver six times as big a payload into orbit as a single-stage system. Subsequent stages do not gain quite so much, and diminishing returns become apparent. Additional drawbacks to multiple stages include cost (e.g. extra rocket motors in the launcher) and reliability. Each additional stage introduces complexity and the possibility of failure from separation problems or the overall reliability of the rocket motors.[26] All space flights to date except the recently terminated Space Shuttle and SpaceX Falcon have used multi-stage expendable launch systems.

Launch Site: The launch pad is a fixed structure designed to launch airborne vehicles or rockets into space. It generally includes the launch tower and a flame trench or deflector. The launch site, in addition, invariably also houses infrastructure to erect, fuel and maintain the launch vehicles. Practical launch sites need to satisfy various competing requirements. While the rocket initially flies straight up to clear the launch site, it needs to accelerate in the desired flight path soon after launch to clear the atmosphere, with a clear ground track below to guard against mishaps en route.[27] The USA launches its space vehicles over water. The launch sites on its eastern coast are at the Eastern Range at Cape Canaveral Air Force Station/Kennedy Space Centre, Florida, and NASA's Wallops Flight Facility, Virginia, while on its western coast is the Western Range at Vandenberg Air Force Base, California.[28] Russia (and previously the Soviet Union) uses landlocked sites in remote areas of Asia, while the European Space Agency (ESA) with its launch site at Kourou in French Guyana, provides it access to an east facing track over the Atlantic.[29] Israel has a launch site that only permits westward launches while India launches eastwards from Sriharikota over the Bay of Bengal. Amongst many considerations for launch site location, one significant one is the advantage gained by an east-facing launch. It has already been seen above that a satellite needs to be imparted the required orbital velocity of 29,000 km per hour for an orbit just above the surface of the Earth. The Earth's speed of rotation eastwards around its north-south axis at the Equator, is approximately 1,675 km per hour. Thus, anything launched eastwards, would have the *ab initio* advantage of already possessing this rotational speed and would need an additional speed of 27,325 km per hour to reach the orbital speed of 29,000 km per hour. While the Earth's rotational speed obviously is too minimal, it still is a significant gain as it saves energy and allows for larger payloads, and an advantage favouring launch sites near the Equator (where the Earth's rotational speed is maximum) for many orbits.

Launch Window: For a satellite to be launched, certain parameters would need to be met. Different mission-based launches would need to meet different parameters. A launch for say, Mars, would have to address the issue of when and where Mars come closest to the Earth. A satellite requiring solar power early in its mission would, thus, be constrained to launch into the sunlit part of the orbit. A spacecraft heading to rendezvous with the Space Station would need to be timed to account for the orbital motion of the Space Station itself. Thus, each of the above launches would have a different launch window, unique to its mission.

Launch Detection: There is a variety of overt indicators of potential space activity. A nation may publicly announce its intentions to launch for reasons of prestige or influence. There may be open announcements regarding assembling of the payload or launcher, and space reconnaissance may reveal related activity at the launch site. Most of these indicators could be circumvented if it so suited that nation's interest. The first unambiguous indicator of space activity is the launch itself.[30] The quest for surveillance of space launches and space-based assets, owes its origin largely to the days of the Cold War – an era that witnessed a tremendous pace of activity by the two superpowers in the *race for space*. Besides the international and national prestige associated with launching the first human in space and the first human on the Moon, equal frenzy was seen in the fields of weaponisation of space. Ballistic missiles and defensive measures to detect and then to counter them saw equal focus, if not more. Amongst various means available to detect such launches, the radar, acronym for radio detection and ranging, stood at the forefront. Most radars are line-of-sight systems, though the ballistic missile threat precipitated the need for over-the-horizon capability. Besides radar, there is heavy reliance on Infra-Red (IR) detection. IR radiation, the band of electromagnetic radiation with wavelength longer than that of visible light, also commonly referred to as heat radiation, is emitted by all bodies, natural and man-made, with temperature above *absolute zero*, i.e., -273.15 degree Celsius. The IR signature given out by a rocket's exhaust is distinctive and is concentrated around specific IR wavelengths. IR detectors can even differentiate between solid and liquid fuelled devices. Once the launched object achieves orbit, the two main methods of tracking it are via radar and optical systems.[31]

Deorbit and Reentry: There are eventualities when an object in orbit is required to be intentionally deorbited. The orbital slot may be required for reuse by a replacement. A satellite may pose a hazard if for some reason it reenters in an uncontrolled manner and the debris control will necessitate an intentional deorbit into a safe area. To prevent a satellite from breaking up either due to collision or structural failure due to age, deorbit mitigates against the resulting debris hazard in space. In all these cases, there is the need to reduce the speed of the satellite, lower its orbit and make it reenter the Earth's atmosphere. The deorbit process is started by firing a retrograde rocket (firing in the opposite direction), or retro firing, to reduce the forward velocity of the satellite. As was seen above in the concept of orbital motion, as the speed of the satellite is reduced, it drops into a lower orbit. The initial deceleration is, thus, provided by retro firing. As the satellite enters lower altitude orbits, atmospheric drag comes into play. Assuming the satellite is not designed to survive reentry, much of the structure will burn up or break down into smaller pieces in the upper atmosphere.[32] For some satellites, there still exists the possibility that sizeable pieces may impact the Earth. To prevent damage from the resulting debris, controlled reentry must cater for safe impact areas, including large water bodies. Manned space missions aboard Apollo provided heat shielding and controlled reentry for sea recovery, while Space Shuttle missions using reusable spacecraft have landed on runways.

Control of Space Assets

The magic is on the ground.

— Col Edward T. Cope
Deputy Chief Information Officer
US National Reconnaissance Office

A generic space system infrastructure consists largely of three segments: the **space segment** which includes primary mission satellites, payloads and relay satellites, the **launch segment** which includes the launch vehicles and launch sites, and the **ground segment**. There could be another segment that could be considered, that is, the user segment comprising the end users, user terminals and associated services. Having had a brief look at the first two segments, we take a peek at an equally important component that encompasses the hardware, software and infrastructure that contributes to space mission accomplishment.

Satellites are the news-grabbing visible end of space operations. It is the ground-end without which there could be no exploitation of the space asset. The ground segment supports the launch infrastructure during launch and when the satellite is positioned in orbit, this segment helps in keeping it there.[33] Once in orbit, the satellite's position needs to be monitored. Necessary commands need to be given to manoeuvre the satellite if it deviates or the mission requirements change. Onboard support systems also need monitoring and readjusting in respect of communications, power generation and cooling. Constant monitoring and readjustments are equally true for the onboard payload, wherein onboard sensors may need to be aimed and operated, antennae steered and some payloads may warrant ejection for recovery.[34] The ground system, thus, monitors and controls the mission from launch, to orbit, to operations to mission accomplishment or deactivation. The primary functions of the ground system are three-fold:[35]

- *Mission management* focusses on planning and scheduling of mission activities, payload tasking, evaluation of system performance and real-time reactions to events and changing circumstances.
- *Asset command and control* includes all activities and functions that enable a spacecraft or payload to perform its mission through its life, relying heavily on the ground-space asset-ground communications link. A major aspect within this is the Telemetry, Tracking and Commanding, or TT&C. Telemetry is the transmission of automated health and status measurements from the spacecraft to ground control, while tracking (or orbit determination) is the determination of orbital parameters based on angular position, range and range-rate measurements and commanding refers to transmissions from the ground to the spacecraft to affect operations and reconfigure payloads and subsystems.
- *Mission data processing* includes all functions and activities related to the information gathered in space, from the time it arrives on the ground from the space object till the processed data, communications or product reaches the end users. It would be evident that the ground systems are inter-related and complex, and a decision affecting one component will generally affect others within the system. Perhaps the most flexible and accessible system within the space system architecture, the ground segment or system is also versatile enough to be called upon to accommodate unplanned changes or errors in space segment design or operation.

Considerations for Manned Space Flight: While the theory behind launch of a manned spacecraft would remain essentially the same as that of an unmanned one, certain additional factors come into play, the paramount among them being crew safety. Not only is the standard of reliability for manned flight greatly enhanced with respect to individual component functioning to cater for all anticipated contingencies and emergencies, adequate escape mechanisms need to be built in to allow the crew to escape in case of beyond-control failures. The human crew, of course, requires life support systems to provide habitable accommodation throughout the flight and safe reentry at the end of the mission. This includes air conditioning, pressurisation, launch and recovery constraints on allowable acceleration, and provisioning of food and water. All such measures come with the added price of payload design and payload mass.[36] Apart from the relatively short duration Apollo missions, manned space activity has been restricted to relatively low altitudes of below 500 km, where the Earth's geomagnetic field accords protection against harmful radiation. Radiation shielding required in higher altitudes, where the Earth's geomagnetic field's protection is weaker, would make sustained occupation by a manned spacecraft problematic.

Spacecraft for Manned Space Flight:[37] Spacecraft that is designed for human space travellers provides for safe travel in vacuum conditions, extremes of temperatures and radiation hazards of space, right from launch to landing. The life support system must reliably provide the crew with all their daily needs of clean air, potable water, food and reliable waste removal. The food, as per individual member's choice of menu, is prepared and along with water, stocked in the module prior to launch. The space food trays have magnets, clamps and double adhesive tape to hold metal, plastic and other utensils. Forks and knives can be used in the micro-gravity conditions, but members learn to eat without sudden starts and stops, so their meal does not float away. The spacecraft is designed around a pressurised cabin or module so that each member of the crew has sufficient room to work and live in 'shirtsleeve' comfort, while individual space suits are worn only to perform extravehicular operations. Each member has enough 'personal space' to work, eat, relax and sleep. The air revitalisation sub-system maintains a breathable, comfortable atmosphere of the proper gas composition (80 percent nitrogen and 20 percent oxygen in US manned spacecraft), temperature, humidity and sea-level pressure. The air-handling sub-system also controls airborne contaminants of all kinds. The waste-management sub-system handles, treats and disposes human by-products, both solid and liquid. Because of the growing space debris risk, trash and solid waste products are stored in containers and brought back to the Earth, while liquid waste products are ejected into space.

Orbital Concepts

Satellites may also be classified by the height at which they orbit the Earth. Different altitude bands offer different opportunities for use of space. Just as different seats in a theatre provide different perspectives on a performance, different Earth orbits give satellites varying perspectives, each valuable for different reasons. Some seem to hover over a single spot, providing a constant view of one face of the Earth, while others circle the planet zipping over many different places in a day. The height of the orbit determines how quickly the satellite moves around the Earth. An Earth-orbiting satellite's motion is mostly controlled by the Earth's gravity. As satellites get closer to Earth, the pull of gravity gets stronger, and the satellite needs to move more quickly to stay in orbit. NASA's Aqua satellite, for example, takes 99 minutes to orbit the Earth at about 705 km altitude, while a weather satellite about 36,000 km from the Earth's surface takes 24 hours, to complete an orbit.[38]

Going by the altitude classification, there are essentially three types of Earth orbits: Low Earth Orbit (LEO), Medium Earth Orbit (MEO), and High Earth Orbit (HEO). Most scientific satellites and Earth observing systems have a LEO. Satellites that operate in MEO include navigation and specialty satellites, designed to monitor a particular region. HEO has applications for weather, navigation and communications.

Low Earth Orbit

LEO is an orbit from approximately 160 to 2,000 km[39] above the Earth's surface, while British standards define this band from 100 to 1,200 km.[40] Most of all human space flights, remote sensing, scientific satellites and many weather satellites operate in the region of LEO, as also flights of the (now terminated) Space Shuttle and the orbit of the International Space Station (ISS).

Even though the Earth's atmosphere at this altitude is extremely thin, there is a significant amount of drag exerted on objects within a LEO depending on their altitude, requiring them to expend some energy to remain in orbit. A satellite orbiting at the lower end spectrum of LEO could stay in orbit for only a few days before decaying back while one at the higher end of LEO will remain in orbit for years or even decades, before needing a boost. Because this drag increases as altitude diminishes, it is not common for satellites to orbit at less than around 300 km altitude.

Satellites operating in LEO travel around the world in 90-120 minutes at a speed of about 8 km per second. Launching a satellite into LEO takes much less energy than launching it above LEO and the equipment needed to send a signal back to the Earth can be much less powerful. It is in this band that trade-offs between drag, area coverage and optical resolution become necessary. Being close to the Earth's surface, LEO offers the ideal altitude for satellites designed for the optical reconnaissance role. In fact, the most valuable aspect of LEO is its proximity to the Earth for observation and low-powered communications. LEO satellites, in addition, are relatively easier to launch, with heavier payloads. For these reasons, LEO satellites are widely used despite the fact they cannot remain situated over one part of the planet and also that LEO satellites are most vulnerable to man-made orbital debris. Operating closer to the Earth's surface, satellites typically have fast overflight times and narrower horizon-look or perspective, commonly called the footprint. Being in the 'field of vision' of a ground control station for a shorter duration, LEO satellites depend on multiple ground stations or satellite relays for control and communications.

A number of different manned activities also take place in LEO. The most notable of these is the International Space Station, which operates at an altitude of around 320 to 347 km. The International Space Station is replenished regularly by the Soyuz spacecraft and the recently terminated Space Shuttle, all of which operate only in LEO missions specifically below 500 km altitude.[41] Significant protection accorded by the Earth's geomagnetic field to space travellers in LEO from cosmic rays and solar flare environment is not the only reason. The amount of energy needed to place a vehicle outside LEO is enormous, and returning can be tricky, making manned flights particularly challenging. The Apollo programme, which eventually sent men to the Moon, is the only known programme to send humans outside of LEO.

Sun-synchronous orbit, though not altitude-defined, operates in the LEO band and is unique in its application, in that it describes the orbit of a satellite that provides a consistent view of the Earth. The satellite in such an orbit keeps a fixed orientation with respect to the Sun and passes the Equator at the same local time each day.[42] This consistent view of that part of the Earth offers tremendous advantage to satellites that have a role towards weather monitoring, remote sensing (those carrying ocean and atmospheric remote sensing instruments that require sunlight) and keeping that tract of the Earth's surface under regular surveillance.

Medium Earth Orbit

One historical definition of Medium Earth Orbit (MEO) that finds most common use is that it lies between the top band of LEO and just below 36,000 km. In practice many MEO satellites operate in a circular orbit at 20,200 km so that their period is exactly 12 hours. This means that they complete exactly two orbits in a day or one orbit in half a day, giving the orbit its alternative name of a "semi-synchronous orbit".[43]

NASA explains the MEO band as comprising essentially two notable orbits: the semi-synchronous orbit as above and the Molniya orbit. Invented by the Russians, the Molniya orbit works well for observing Earth regions at high latitudes. A geostationary orbit, being discussed subsequently under High Earth Orbit (HEO), is valuable for the constant view it provides but satellites in a geostationary orbit are parked over the Equator and don't work well for the far northern or southern regions of the Earth which are always on the edge of view of a geostationary satellite. Thus, large parts of the erstwhile mainland Soviet Union remain outside or on the fringes of the footprint of a geostationary satellite and, thus, arose the need of a 'different kind of orbit' to address this specific concern of the Soviets. The Molniya orbit offers a useful alternative. The Molniya orbit is highly inclined, that is, the plane of the orbit is inclined away from the plane of Equator. The satellite moves in an extreme ellipse or eccentric orbit with the Earth close to one edge of the ellipse. Because it is accelerated by the planet's gravity, the satellite moves very quickly when it is close to the Earth. In fact, the word "Molniya" is the Russian word for lightning, to describe the speed at which a satellite in this particular orbit travels through its closest point of approach. As it moves away, its speed slows, so it spends more time at the top of its orbit, farthest from the Earth. A satellite in a Molniya orbit takes 12 hours to complete its orbit, but it spends about two-thirds of that time over the hemisphere of interest.[44]

Fig 1.2: The Molniya Orbit

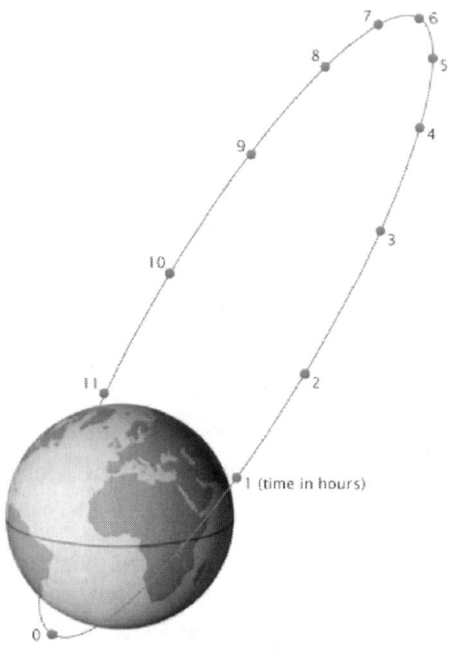

Adapted from http://earthobservatory.nasa.gov/Features/OrbitsCatalog/page2.php.

In fact, Molniya orbits are a special case of semi-synchronous orbit wherein each satellite in this orbit is usable for eight hours in a 12-hour orbit. The highly elliptical 12-hour orbit places the highest point in the orbit, the apogee, at 40,000 km over the northern hemisphere and the lowest point, the perigee, about 500 km over the southern hemisphere.[45] While 'semi-synchronousity' is the common factor between the Molniya orbit and MEO, as is being discussed here under MEO, it could also well be argued that the Molniya orbit is a HEO[46] as its apogee touches 40,000 km.

Like LEOs, MEO satellites don't maintain a stationary distance from the Earth. While about two dozen LEOs would be required to provide continuous coverage, the same could be achieved with fewer MEOs due to their higher orbit altitudes. However, LEOs typically orbit in a circular pattern around the Equator. A MEO satellite may have a variety of different orbits, including elliptical ones and may provide better overall coverage of satellite communications if enough of them are in place and the orbit is swift. MEOs typically are able to create a larger footprint because of their different orbital patterns and because they are higher than LEOs. The high-radiation hostile environment of the Van Allen belts makes the MEO region inhospitable to long duration manned flights[47] besides creating hazards to electronic systems.

Today, MEO is most commonly used in satellite-based navigation systems around the world. These include the US' Global Positioning System (GPS) and the Russian Glonass.

High Earth Orbit

While LEO satellites have a typical orbital period of 90 to 120 minutes, semi-synchronous MEO satellites have a period of 12 hours. As the altitude of operation is increased, the period of orbit increases. At an altitude of 35,800 km,[48] the orbital period is 24 hours, which is to say, the Earth and the satellite rotate at the same rate. Satellites in such an orbit are called geosynchronous satellites. Any orbit at this altitude will have a 24-hour period, regardless of its inclination. The orbit could be over the Equator, over the poles (polar orbit) or anywhere in between.

The geosynchronous Earth orbit over the Equator, however, is unique, in that it appears to be stationary to an observer on the Earth. A satellite in a circular geosynchronous west-to-east orbit directly over the Equator (with zero eccentricity and inclination) will have an orbit that does not move at all relative to the ground, as the Earth too is rotating in the same direction. It is always directly above and 'parked' over the same point on the Earth's surface. Such an orbit is not only geosynchronous but also geostationary and termed the Geostationary Earth Orbit (GEO). Specifying the fixed point on the Equator defines the location of a geostationary satellite, so there can be only one geostationary orbit for every longitude. Fortunately, since the circumference of the geostationary orbit is 267,000 km, there is room for quite a few satellites. From this altitude above the Earth's surface, the geostationary satellite can see almost one-third of the Earth's surface. Not only do satellites in such orbits have a larger footprint, they also make excellent platforms for communications relays, nuclear detection, missile warning and surveillance missions. This orbit is extremely valuable for weather monitoring because satellites in this orbit provide a constant view of the same surface area. A relatively small number of satellites in such an orbit is required to provide global coverage.[49]

Under 'Launch Considerations' discussed earlier in this chapter, we saw the need to intentionally deorbit a no-longer-useful space object by controlling its reentry back to the Earth. For a satellite as high as in HEO, a substantial amount of propellant would need to be catered for to safely deorbit and reenter the satellite at the end of its operational life. Carriage of this additional amount of propellant would be at the cost of the propellant required for the mission of the satellite and would impinge upon the satellite's operational life. A practical alternative usually undertaken is to save a smaller amount of propellant, to boost the satellite just beyond the HEO altitude. This action creates a 'space graveyard' above HEO, while freeing an orbital slot in the much valued GEO belt. Orbital decay from such a graveyard orbit is genuinely negligible[50] as modelling indicates decay time predictions of millions of years.

Beyond geosynchronous altitude, satellites travel slower relative to the Earth and appear to regress through space. Although very few satellites operate above the geosynchronous altitude, these altitudes do offer advantages for increasing satellite survivability.

Fig 1.3: Earth Orbits

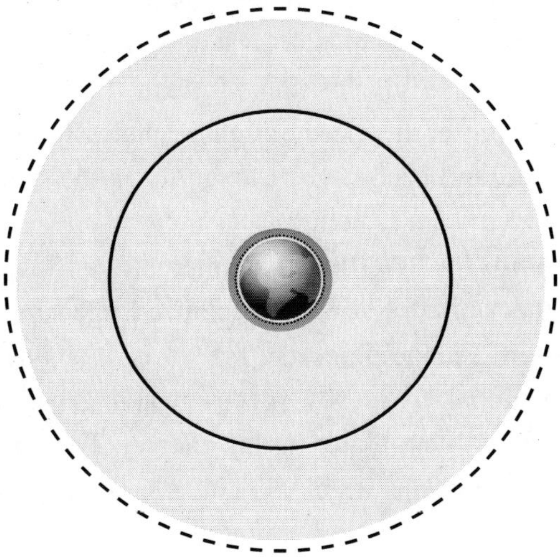

Various Earth orbits to scale; innermost, the dotted line represents the orbit of the ISS; the dark grey represents LEO, the light grey represents MEO, black line represents the semi-synchronous orbit and the dashed line represents the geosynchronous orbit.

Sourced from http://en.wikipedia.org/wiki/List_of_orbits, on January 4, 2012.

Communicating with Satellites

The use of satellites has revolutionised the field of communications and satellite-based services worldwide, in both the commercial and military domains. The efficiency of a satellite's service, however, remains defined by the strength of the satellite link, the link between the Earth station and the satellite. The signal or command from the Earth to the satellite is conveyed on the *uplink frequency,* which needs to be only strong enough to reach the addressed satellite, after, of course, catering for atmospheric attenuation. At the satellite end, the signal is processed to remove noise and analysed to extract routing information.[51] The signal then is rebroadcast back to the Earth on a different *downlink frequency*. Use of two separate frequencies for uplink and downlink allows the uplink receiver and the downlink transmitter in the satellite to operate simultaneously without mutual interference. Most satellites have multiple numbers of such receiver/transmitter pairs, called transponders. As the downlink transmitter has to rely solely on the power system of the satellite, there are greater constraints on downlink signal power than for the Earth station. While this signal from the Earth to the satellite, and back to the Earth, travels at the speed of light, technically speaking, it takes a finite amount of time, and needs to be catered for in the specific application of the service provided by the satellite.[52]

Frequency Bands for Satellites: The International Telecommunication Union (ITU) is the United Nations' specialised agency for Information and Communication Technologies (ICTs). It is this international body that allocates the global frequency spectrum and geostationary satellite orbits and is covered in detail later in this chapter. There being over three billion frequencies in the spectrum[53], ITU has produced a comprehensive and relatively simple classification of frequency bands, without any gaps or overlaps between bands, across the entire spectrum from Extremely Low Frequency (ELF) band (3-30 Hz) to Extremely High Frequency (EHF) band (30-300 GHz). Various bands offer various capabilities and functionalities across the entire breadth of communication activity right from underwater, terrestrial, air, space to outer space. Communication frequencies for satellites are mainly in the Super High Frequency (SHF) band (3-30 GHz), while the Ultra High Frequency (UHF) band (300-3000 MHz) and EHF band are also used. The American Institute of Electrical and Electronic Engineers (IEEE) further sub-divides the UHF, SHF and EHF bands, denoting the sub-bands by alphabets. IEEE (Specification 521-2002) defines these bands as:[54]

Table 1.1

Band	Frequency	Services[55]
L	1-2 GHz	Mobile and sea communication, messaging
S	2-4 GHz	-do-
C	4-8 GHz	Commercial communications
X	8-12 GHz	Reserved by the US military
Ku	12-18 GHz	Communication between commercial users
K	18-26 GHz	Reserved by the US military
Ka	26-40 GHz	Broadband communications

Frequency and Orbit Allocation: Despite the availability of a very wide frequency spectrum, usable frequency segments are vied for by terrestrial radio networks, wireless networks and military communications. Into this myriad are added the communications for satellites. Unlike the terrestrial based communications which largely operate on the *line of sight* principle and are mostly limited to 'range to horizon', satellite communications are unique in their own way. Communication satellites have the ability to not only cause interference and receive interference from terrestrial networks, but also from neighbouring satellites operating on the same frequencies. While the severity of such interference from terrestrial sources is reduced for satellites in Geostationary Earth Orbit (GEO) due to their large altitude above the horizon, GEO satellites typically operating in the 7-8 GHz band,[56] face an additional issue of sheer numbers. With a sizeable number of them already in operation in GEO, the complexity increases as more and more of them are being launched and planned for future launches. While the 7-8 GHz band remains limited, frequencies on offer become less and less and orbital separation gets closer and closer, leading to a potential vulnerability of unintentional inter-satellite interference or jamming. It is here that ITU steps in to administer conflict resolution amongst satellite operators.

ITU: The ITU was founded in Paris in 1865 as the International Telegraph Union. It took its present name in 1934, and in 1947, became a specialised agency of the United Nations. Although its first area of expertise was the telegraph, the work of ITU now covers the whole ICT sector, from digital broadcasting to the Internet, and from mobile technologies to 3D TV. The work holds equally true in allocation and coordination of frequencies for communications, Earth observation and science spacecraft, and any other spacefaring craft that communicates using radio frequencies. An organisation of public-private partnership since its inception, ITU currently has a membership of 193 countries and some 700 private-sector entities.[57] ITU is headquartered in Geneva, Switzerland, and has 12 regional and area offices around the world. Under the auspices of the United Nations, ITU administers the radio frequency spectrum worldwide, in such a manner that the needs of all member nations and agencies are met, in terms of avoidance of interference, and resolution of potential disputes and conflicts. Towards this end, ITU has divided the world into three ITU regions for the purposes of managing the global frequency spectrum. Each region has its own set of allotted frequencies.

Fig 1.4: The Three ITU Regions

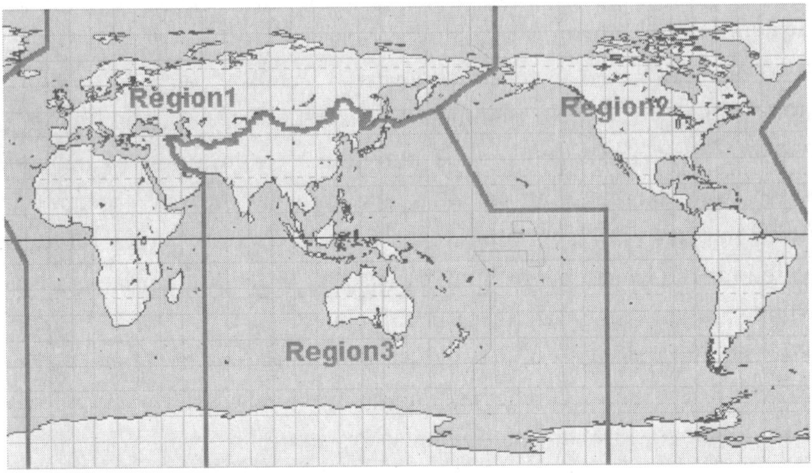

Adapted from http://www4.plala.or.jp/nomrax/ITU_Reg.htm on January 11, 2012.

The aim is not only to see the frequency spectrum as a limited natural resource that must be used rationally, efficiently and economically, but also that countries and groups of countries have equitable access to it. An equally important role carried out by ITU is of allotting GEO orbital slots. All member satellite operators are mandated to register with ITU for both orbital position and frequency before launching the GEO satellite. However, military frequencies are exempt from this regulation other than the requirement to ensure that the military frequencies in use do not cause interference to other users.[58] Once an allocation has been made by ITU, the member state has a fixed time to occupy the allocated orbital position, as failure to do so would invite reallocation to another bidder.[59] Once registered, the cleared slot and the frequency are deemed to be in effect until the operator-defined system's life expectancy has expired, or is no longer in use under due notification to ITU. ITU has, by design, no enforcement powers or authority. The member states choose voluntarily to abide by the rules and regulations enunciated by ITU and work to resolve any conflict about spectrum usage in good faith and with consideration of mutual benefit.[60] ITU remains a coordinating agency for the international use of the spectrum, while the authority to launch and place the satellite in orbit to operate on the given frequencies, rests with the concerned sovereign state. At the sovereign or member state level, the state's administration is the link between ITU and the satellite operator. The state administration, as the domestic regulatory body, participates in the ITU activities and oversees the assignment and use of frequencies within the state's domestic borders.[61] At the planned user end, the user proposing to launch a GEO satellite applies for the orbit and frequency to the domestic regulatory body, which, in turn, conducts an internal coordination and applies it to the ITU five to two years prior to the launch of the operation, specifying details of channel frequencies, orbital position, geographical location of the intended Earth station, transmitter power and data format. On due approval from the ITU, the member state is now ready to launch its satellite.

Notes

1. Erika Cheetham, *The Final Prophecies of Nostradamus* (London: Warner Books, 1993), p. 480.
2. Development, Concepts and Doctrine Centre (DCDC), Ministry of Defence, UK, *The UK Military Space Primer* (Wiltshire: DCDC, 2010), Ch. 1, pp. 1-2.
3. Ibid., Ch. 2, p. 2.
4. The line was named after Theodore von Kármán, (1881–1963) a Hungarian-American engineer and physicist who was active primarily in the fields of aeronautics and astronautics. He first calculated that around this altitude, the Earth's atmosphere becomes too thin for aeronautical purposes because any vehicle at this altitude would have to travel faster than orbital velocity in order to derive sufficient aerodynamic lift from the atmosphere to support itself.
5. DCDC, n. 2, Ch. 1, p. 2.
6. Joseph A. Angelo, Jr, *Encyclopedia of Space and Astronomy* (New Delhi: Viva Books Private Limited, 2010), p. 554.
7. Ibid., p. 564.
8. Ibid.
9. DCDC, n. 2, Ch. 1, p. 6.
10. Ibid.
11. Angelo, n. 6, p. 218.
12. It is a misconception that satellites remain in orbit because they have exited the Earth's gravity. Satellites are held in orbit by the gravity of the planet which they are orbiting. The fallacy is further compounded by almost universal misuse of the

term "zero gravity" to describe the free-falling conditions aboard orbiting space vehicles. Gravity still exists in space. It keeps satellites from flying straight off into deep space. What's missing is "weight", the resistance of gravitational attraction of the planet. Satellites stay in space because of their tremendous horizontal speed, which allows them — while being unavoidably pulled toward the Earth by gravity — to go "over the horizon." The ground's curved withdrawal along the Earth's round surface offsets the satellites' fall towards the ground. Speed and not lack of gravity keeps satellites in orbit around the Earth. Because the inertial trajectory of a satellite or a spacecraft compensates for the force of the Earth's gravity, an orbiting platform and all its contents approach a state of free fall. In this state, all objects and space travellers aboard the spacecraft appear "weightless".

13. DCDC, n. 2, Ch. 1, p. 8.
14. Angelo, n. 6, p. 558
15. DCDC, n. 2, Ch. 1, p. 10.
16. Joseph A. Angelo, Jr, *Space Technology* (New Delhi: Pentagon Press, 2009), p. 6.
17. Ibid., p. 4.
18. DCDC, n. 2, Ch. 1, p. 24.
19. Ibid, Ch. 1, p. 25.
20 http://space.au.af.mil/au-18-2009/au-18_chap06.pdf, accessed on December 14, 2011.
21. DCDC, n. 2, Ch. 1, p. 37.
22. Angelo, n. 16, p. 121.
23. Angelo, n. 6, p. 510.
24. DCDC, n. 2, Ch. 1, p. 55.
25. The Indian Space Research Organisation's forays into space have banked upon the four-stage solid-fuelled Satellite Launch Vehicle (SLV), five-stage solid-fuelled Augmented SLV and four-stage alternate solid- and liquid-fuelled Polar SLV.
26. DCDC, n. 2, Ch. 1, p. 59.
27. Ibid., Ch. 1, p. 59.
28. Angelo, n. 6, p. 349.
29. DCDC, n. 2, Ch. 1, p. 59.
30. Ibid.
31. DCDC, n. 2, Ch. 3, p. 57.
32. Ibid., Ch 7, p. 72.
33. http://www.aero.org/publications/crosslink/pdfs/V7N1.pdf, accessed on December 21, 2011.
34. DCDC, n. 2, Ch. 1, p. 77.
35. http://www.aero.org/publications/crosslink/pdfs/V7N1.pdf, accessed on December 21, 2011

36. DCDC, n. 2, Ch. 4, p. 11.
37. Angelo, n. 16, pp. 175-7.
38. http://earthobservatory.nasa.gov/Features/OrbitsCatalog/, accessed on November 30, 2011.
39. Ibid.
40. DCDC, n. 2, Ch. 1, p. 46.
41. Angelo, n. 6, p. 565.
42. Usually, such orbits are described by the local time at one of the points of equatorial crossing. DCDC, n. 2, Ch. 1, p. 51.
43. Ibid., Ch. 1, p. 47.
44. http://earthobservatory.nasa.gov/Features/OrbitsCatalog/page2.php, accessed on January 8, 2012.
45. Angelo, n. 6, pp. 405-406.
46. DCDC, n. 2, Ch. 1, p. 50.
47. The manned orbital altitude record was set by the crew of the US Gemini XI mission in 1966. They flew in an elliptical orbit with a maximum altitude of 1,370 km and a minimum altitude of 160 km; so technically, they intruded briefly into what might be regarded as MEO. Their mission lasted just under three days.
48. DCDC, n. 2, Ch. 1, p. 48.
49. Full global coverage between 70° North latitude and 70° South latitude can theoretically be achieved using three equally spaced (120° apart) geostationary satellites. Operationally, four or more satellites are required to provide this coverage in order to mitigate the effects of a satellite failure on the networks. Four satellites provide overlapping capabilities, greater traffic handling capacity and a measure of redundancy. http://space.au.af.mil/primer/satellite_communications.pdf, accessed on January 9, 2012.
50. DCDC, n. 2, Ch. 1, p. 71.
51. Ibid., Ch. 3, p. 10.
52. The speed of light is 300,000 km per second, or $3x10^{8}$ m per second. This may appear unbelievably fast, but it still takes light from the Sun approximately eight minutes to reach the Earth. A signal from the Earth to a satellite in GEO and back to the Earth, thus, takes about one-fourth of a second to travel.
53. Wg. Cdr. Kaza Lalitendra, *Militarisation of Space* (New Delhi: KW Publishers Pvt. Ltd, 2010), p. 44.
54. DCDC, n. 2, Ch. 3, p. 13.
55. Cesar Jaramillo, ed., "Space Security Index 2012," accessed at http://www.spacesecurity.org/SpaceSecurityReport2012.pdf on May 15, 2013. Besides the X and K bands, the UHF band (240-340 MHz) too has tremendous utility for the US military.

56. Ram Jakhu and Karan Singh, "Space Security and Competition for Radio Frequencies and Geostationary Slots", Paper presented at the Space Security Conference at the Institute for Defence Studies and Analysis, New Delhi, on November 13 and 14, 2007.
57. http://www.itu.int/en/about/Pages/history.aspx, accessed on January 10, 2012.
58. Lalitendra, n. 53, p. 44.
59. DCDC, n. 2, Ch. 2, p. 7.
60. Sqn. Ldr. K.K. Nair, *Space: The Frontiers of Modern Defence* (New Delhi: Knowledge World, 2006), p. 104
61. Bruce R. Elbert, *The Satellite Communication Applications Handbook* (Boston, USA: Artech House, Inc, 1997), Ch. 10, p. 430.

2

Dawn of the Space Age

Mankind will not remain tied to the Earth forever.
— Inscription on the tombstone of Konstantin Tsiolkovsky

The Need for High Ground

A mention was made in the previous chapter of the evolution of civilisation in the context of the three dimensions. As man began exploring in his quest for conquering land, he took to capturing *high ground* for the inherent tactical advantages that accrued. Not only was the acquisition of *high ground* a necessity for dominating the area around it but it gave the conqueror a higher platform for surveillance. This *high ground* gave him the capability of observation, providing him early warning of the threats that could manifest themselves. As the art of war-fighting evolved, he built his defences on this ground of tactical significance and to enable early warning, he extended his *line of sight* or reach of observation by raising the platforms of observation and positioning them beyond his defences. It was, thus, only a matter of time that man would look to the skies and seek to place a platform in the air.

The first recorded airborne observation post for military use was the employment of the hot air balloon by the French against Austria in the Battle of Fleurus, 1794.[1] The French triumphed and the Austrians protested that the use of a balloon was against the rules of war. Hot air balloons continued to be used for airborne reconnaissance and saw active participation in the American Civil War (1861-65) by the Union Army, Franco-Prussian War (1870-71) by the French, Eritrean War (1887-88) by the Italians, and the Boer War in South Africa in 1900 by the British Army, amongst others.[2]

December 17, 1903, was the day that changed the way the third dimension was to be used thereafter. This was the day that witnessed the world's first successful airplane making the first controlled, powered and sustained heavier-than-air human flight. The Wright Brothers had arrived and heralded a new game-changing era in aviation history. World War I ushered in the military applications of fixed-wing aircraft, in the roles of reconnaissance and ground attack. Hot air balloons continued to be employed as reconnaissance platforms of the *high ground* and were found to be valuable targets by the enemy aircraft.

The post World War I era saw an exponential increase in technology, resulting in extreme versatility in the role played by air power during World War II. The *high ground* assumed equal, if not more, importance, in that the war saw a plethora of dedicated aircraft specifically designed for strategic and tactical reconnaissance.

It was thereafter the age of the Cold War that moved the *high ground* higher. The medium that provided space for air-dependent aero-engines made way for the medium that employed air-independent rocket engines, enabling spacecraft to operate in regions outside the influence of land and air wars. It was the quest for the first human in space and the first on the Moon, that set the pace between the two superpowers, ushering in the space race. It was a product of this technology driven race that saw immense innovation and scientific development by the two superpowers, enabling mastery over the technology of rockets, launch vehicles and space payloads, taking mankind's forays higher and further.

Rocketry in Antiquity

Any endeavour to reach space is centred around the rocket which provides the power to propel the spacecraft into space. When and how rockets appeared in the annals of human civilisation remains unrecorded, though the Chinese are credited with the use of gunpowder primarily for fireworks during their festivals, as early as the first century C.E.[3] Bamboo tubes filled with gunpowder were tossed into a fire. Perhaps one of the tubes was improperly sealed and did not explode but went skittering out of the fire, propelled by the exhaust of the hot gases from the burning gunpowder mixture. It could be conjectured that this phenomenon would have been observed and thereafter taken forward by experiments. At some stage, the early rocketeer attached the bamboo tube to an arrow and launched it with a bow, and it is likely this "fire arrow" that finds a mention in numerous Chinese documents. It was, however, in 1232 AD that the use of rockets in warfare was first found reported.[4] The Chinese repelled the Mongolian invaders in the Battle of K'ai-fung-fu, with a barrage of "flaming arrows". The rockets were huge and apparently quite powerful. According to a report, "When the rocket was lit, it made a noise that resembled thunder that could be heard for five leagues -- about 15 miles. When it fell to the Earth, the point of impact was devastated for 2,000 feet in all directions." Apparently these large military rockets carried incendiary material and iron shrapnel.[5]

The Mongolians were quick to adopt this weaponry and spread rocket technology as they invaded the Middle East and Europe, and by the 14th century, gunpowder-rocket technology had proliferated throughout Western Europe.[6] While rockets continued to be used thereafter in warfare, the rocket's potential largely went unrecognised as militaries relied more on the increasingly accurate cannon. Major advances in rocket technology in the Western world did not occur until the early 1900s.[7]

It would be interesting to note that while rocket technology was largely being used for firework displays in Western Europe, especially in the 18th century, it was exploited as a successful weapon of war in the Indian subcontinent.[8] Tipu Sultan (1750-99), the Muslim ruler of the Kingdom of Mysore, championed the use of mass attacks with rocket brigades within his Army. He successfully used metal-cylinder rockets against the larger forces of the British East India Company in the two Battles of Srirangapattanam in 1792 and 1799. Profiting from their Indian experience, the British transported a large number of these rockets to England[9] and led by Sir William Congreve, developed a series of barrage rockets ranging in weight from 8 to 136 kg, which were also used against Napoleon's French Army.[10] His pioneering work on these early solid propellant rockets was a significant technical step in the overall evolution of the modern military rocket.[11] As these rockets were largely inaccurate, reliance was placed on their use in large numbers for causing destruction. It was in 1844 that accuracy was improved by providing spin stabilisation to the rocket, by an English inventor, William Hale. By the end of the 19th century, rocketry and allied technologies had migrated from the East to the West, where these continued to be refined and improved.[12]

Heralding the 20th Century

At the turn of the century, space flight was still in the realm of science fiction. It was only in the 1930s that men responsible for reaching space took their first step towards that goal. Jules Verne in his 1865 novel, *From the Earth to the Moon*, suggested using a gun for propulsion beyond the atmosphere, an idea that caught the fancy of a young Russian school teacher named **Konstantin Tsiolkovsky** (1857-1935).[13] Regarded as the father of human space flight and one of the founding fathers of modern rocketry, Tsiolkovsky, far ahead of his time, wrote a series of articles about the theory of rocketry and space flight at the turn of the century. He figured out theoretically the basic rocket equation, which was published in a Russian magazine, *The Science Review*, in 1903, the same year that saw the Wright Brothers' first flight in America. In an article titled "Research into Interplanetary Space by Means of Rocket Power", he accurately described the state of weightlessness and the function of rockets in a vacuum. He anticipated and planned for a future when mankind would no longer be confined to live in one world and famously quoted, "The Earth is the cradle of humanity, but one cannot live in the cradle forever".[14]

Amongst other things, Tsiolkovsky's visionary works identified the necessity of liquid-fuelled rockets in the future, the very devices that the American physicist Robert H. Goddard would later develop.[15] Some of his other ideas that have since seen the light of the day in some form or the other, included the concept of zero gravity, space suits, space stations, artificial satellites orbiting the Earth, multi-stage rockets, use of solar energy and closed life support systems.

Tsiolkovsky's work and revolutionary ideas[16] influenced many young scientists and engineers, including Sergei Korolov (1906-66), who would later be credited for building the first rocket to fly a human in space.

Geopolitical circumstances in Czarist Russia in those times prevented other scientists of the world scientific community outside Russia from being aware of Tsiolkovsky's work. Parallel developments in the United States and Germany would challenge scientists in similar fields.

In 1926, Konstantin Tsiolkovsky published a grand "Plan of Space Exploration", which outlined 16 steps for mankind's futuristic exploration into space. These give an insight to his magnificent, far-reaching and prophetically visionary ideas:

- *Creation of rocket airplanes with wings.*
- *Progressively increasing the speed and altitude of these airplanes.*
- *Production of rockets without wings.*
- *Ability to land on the surface of the sea.*
- *Reaching escape velocity and the first flight into Earth orbit.*
- *Lengthening rocket flight times in space.*
- *Experimental use of plants to make an artificial atmosphere in space.*
- *Using pressurised space suits for activity outside of space ships.*
- *Making orbiting greenhouses for plants.*
- *Constructing large orbital habitats around the Earth.*
- *Using solar radiation to grow food, to heat space quarters, and for transport throughout the solar system.*
- *Colonisation of the asteroid belt.*
- *Colonisation of the entire solar system and beyond.*
- *Achievement of individual and social perfection.*
- *Overcrowding of the solar system and the colonisation of the Milky Way (the galaxy).*
- *The Sun begins to die and the people remaining in the solar system's population go to other suns.*

The other two pioneers and founding fathers of this era were Robert H. Goddard, a brilliant physicist of the United States and Hermann J. Oberth, a physicist and rocket engineer from Germany. All three worked independently on the theoretical foundations of rocket propulsion and identified the multi-stage liquid-fuelled rocket as the enabling technology for space exploration.[17] While Tsoilkovsky and Oberth were primarily theoreticians, Goddard went beyond and developed the fundamental technology associated with modern rockets.

Robert H. Goddard (1882-1945), regarded as the father of American rocketry, was also greatly influenced by science fiction writers like Jules Verne and H. G. Wells. His initial formative work commenced with testing of solid-fuelled rockets. One of his experiments in 1915, in a static test experiment in a vacuum chamber, demonstrated that a rocket engine was capable of functioning in a vacuum, allowing for space travel some time in the future. In 1917, when the United States entered World War I, Goddard worked on perfecting rockets as weapons. One of his designs became the technical forerunner of the bazooka, a tube launched anti-tank rocket.[18] It was in 1919 that Goddard summarised his rocket propulsion theory in a report, "A Method of Reaching Extreme Altitudes", the first American scientific work that contained the fundamental principles of rocket propulsion, including the concept of carrying a payload to the Moon. It is ironic that he was ridiculed by the contemporary scientific community and much of the media, for going against the then prevalent grain of thought that a rocket needed air to 'push against'. Goddard chose to work in seclusion thereafter, avoided the media and published as little as possible.[19] On March 26, 1926, Goddard established a ground-breaking milestone and space technology history by launching the world's first liquid-fuelled rocket,[20] and then in 1935, when his rocket exceeded the speed of sound. With 214 patents to his name, he pioneered work on the onboard gyroscopic controlled rocket guidance mechanism, pulse jet engine, turbo pumps for liquid-fuelled engines and liquid-fuelled rocket cluster.

In the same era, visions of space flight and work on rocketry had commenced in Germany too. **Hermann J. Oberth** (1894-1989) was a Romanian-born German (theoretical) rocket engineer who, also inspired by Jules Verne, as a theoretician, helped establish the field of astronautics, paving the way for subsequent German scientists and engineers to design rockets that would have a devastating effect in World War II. Unlike Tsialkovsky and Goddard, Oberth was the one who theorised upon human beings being an integral part of the space travel vision.[21] His publications *The Rocket into Interplanetary Space* in 1923, and *Roads to Space Travel* in 1929, conceptualised major aspects of space travel, including ideas on liquid-fuelled rockets, multi-stage rockets, space navigation, and guided reentry systems. His work inspired Johannes Winkler to form the Society for Space Travel in 1927, the society, which on March 14, 1931, launched Germany's first liquid-fuelled rocket. It was in Germany that pioneering work on such rockets invited tremendous media coverage and caught the attention of the German Army's General Walter Dornberger who too envisioned dreams of space travel and in addition, saw potential military applications. He hired

Wernher von Braun, a young budding rocket engineer, and his team in 1932 and the German Army absorbed the society's efforts and established a large military rocket programme, the German Army Rocket Research Group,[22] which by 1936, was being jointly funded by the German Army and the new German Air Force, the *Luftwaff.*[23] This rocket development team, subsequently established the spawning ground for some of the most significant breakthroughs in rocket technology,[24] at a remote coastal area designated Peenemunde, in the run-up to, and during, World War II,[25] and by 1938, rockets from Peenemunde were already reaching altitudes four times higher than Goddard's in the United States.

Rockets in World War II

As war appeared imminent, militaries of the involved nations provided the impetus for further development in rocketry, and during World War II, produced rockets and missiles of all shapes and sizes. In Germany, however, funding for the rocket programme at Peenemunde experienced the first of many interruptions. Hitler initially decided that the new missiles would not be needed and then decided that they were vital to the war effort. As a result, the soon-to-be dreaded V-2 rocket's production got delayed by about two years. By the beginning of 1940, about $14 million[26] had been spent on the infrastructure and experimentation at Peenemunde, making the German rocket programme the most sophisticated in the world[27] and by 1943, Germany was way ahead in significant advances in rocketry and in the employment of rockets.[28]

The Germans used the V-1 (Vengeance Weapon Number One) and then the V-2 missiles in the latter part of the war. The V-1, commonly called the flying bomb, was a liquid fuelled, pulse-jet unmanned drone aircraft that could carry a 2,000 lb warhead to a range of 370 km. There was no navigation system, so it was simply pointed in the direction of its target. Simple gyrocompasses kept it level and range was controlled by the fuel supply. Although primitive by today's standards, it was the first of what we now call a "cruise missile". The first offensive launch was on June 12, 1943, aimed at London and once the Germans got into their stride, they launched an average of 190 V-1 rockets a day.[29]

The V-1's successor, the V-2 rocket, the 'grandfather' of the large rockets, missiles and space launch vehicles of today, achieved the milestone of being the world's first guided ballistic missile, launching a 2,000 pound warhead up into the stratosphere at 3,500 km per hour, being burnt out, with the warhead continuing on a ballistic trajectory literally falling on the target 200 miles away.[30] With an altitude achieved of 60 miles, this technology marked *mankind's first step into* the *space age.*[31] First launched on October 3, 1942, it was used against targets in Europe and by the end of the war, 1,155 V-2s had been fired against England and another 1,675 against Antwerp and other continental targets.[32] The guidance system for these missiles was imperfect and many did not reach their targets, but they struck without warning (other than the twin sonic boom on reentering the atmosphere just moments before impact) and there was no defence against them.[33] As a result, the V-2s had a terror factor far beyond their capabilities. However, for the Germans, the V-2 offensive came too late to alter the outcome of the war, and by April 1945, the German Army was in full retreat.

While Hitler's suicide on April 30, 1945, signalled the end of the war in Europe, the atomic bombings of Hiroshima and Nagasaki in August 1945 brought the war in the Pacific to a definitive end. The atomic bombings were the result of the emergence of another technology near the end of World War II, with the United States heralding the *atomic age* on July 16, 1945, by successfully testing the world's first atomic bomb in New Mexico. The post World War II era would witness a fast paced and technology-driven competitive race between the United States and the Soviet Union, towards advancements in combination of rocketry and the technology of the atom, that would define the rapidly polarising world order and dominate the geopolitics for decades to come.

Germany's Technology Exodus

The two superpowers with different political ideologies competed for technological dominance in the race for space. What was finally to become the race for space, actually commenced as a race for delivery of the newly demonstrated nuclear weaponry across the intercontinental distances that separated the superpowers. Technological accomplishments by both the superpowers owed their origins to the work and workforce of the German Army Rocket Research Centre at the secret base at Peenemunde. Towards the end of World War II, the Allied victors moved to gain access to the secrets of the V-2.

In Nazi Germany, key personnel feared being assassinated rather than being allowed to fall into the hands of the Allies.[34] Fearing this, von Braun orchestrated the exodus of about 4,000 people from Peenemunde in early 1945. Peenemunde finally fell to the Russians on March 9, 1945, and on May 2, 1945, von Braun's English speaking brother organised the 'surrender' of von Braun's team to the Americans near the Austrian border. Operation Paperclip[35] was set up as a programme by the United States Air Force (USAF) specifically to bring German scientists and intellectuals to America after the war. The Americans simultaneously put together Special Mission 2 in May 1945, to gather and ship a sufficient number of parts to build 100 V-2s in the United States.[36] The underground V-2 production facility at Mittelwerk was captured in April 1945 and was to be turned over to the Russian Army by June 1, 1945. In the nick of time members of Special Mission 2 picked up 100 of everything in sight and shipped the parts to the United States. The experienced workforce and the apparatus necessary for further refinements on V-2s on American soil were, thus, in place, leading to the first American V-2s' successful launch on May 10, 1946.[37]

Post World War II, the British and Soviet governments also gained access to the V-2s' technical designs and the German scientists responsible for creating the rockets through Operation Backfire and Operation Osoaviakhim respectively. In fact, the British were the first to reconstruct and fire the V-2 in October 1945, but were too consumed with post-war reconstruction to make any meaningful progress thereafter. In the case of the Soviet Union, the Red Army, on capturing Peenemunde landed up with most of the hardware but very few scientists and technicians. Stalin moved the remnants of the German rocket team to the Soviet Union in 1946, where the Soviet team took over the primary control of the programme.[38] By the end of 1953, the Russians returned the German rocket team to Germany.

The Route to Space Lay Through Weapons

The cold war would become the great engine, the supreme catalyst, that sent rockets and their cargoes far above Earth and worlds away. If Tsiolkovsky, Oberth, Goddard, and others were the fathers of rocketry, the competition between capitalism and communism was its midwife.

— William E. Burrows

In simple terms, the Cold War can be viewed as an expression of the ideological struggle between Communism and democracy.[39] Post World War II, the future of weapons was to be essentially based on two products of the war: the nuclear weapon and the means of delivering it. The start of the Cold War made the Soviets realise the lead that the Americans had achieved in the former. The Soviets commenced work on both, and placed rocket research activities under the Ministry of Armament, rather than the Ministry of Aviation, resulting in two separate and different industries, working towards heading into space, unlike in the United States where the two were combined into the aerospace industry. The Soviets, with their air force not quite as formidable as that of the US, were quick to realise and capitalise on the potential of the Intercontinental Ballistic Missile (ICBM) for striking over long ranges, and by the end of 1947, had reached a consensus to build an ICBM. Sergei Korolov was appointed to head the Soviets' Department 3, devoted to long-range ballistic missile development, and by April 1949, he had designed the R-14 multi-stage rocket to deliver a 3,000 kg warhead to a range of 2,900 km, a design far more advanced than what von Braun had come up with in the United States.[40]

In the context of the United States, the arrival of the rocket age towards the end of the great war, appeared to challenge the relative safety and protection enjoyed hithertofore by the mainland, separated as it was by the two large oceans and the sheer distances from continental Europe and the Soviet Union. At the start of the post World War II era, the United States possessed the most powerful bomber force in the world[41] and also emerged as the only nuclear capable nation. Seemingly content for the time being, the Americans worked towards refinements in the V-2 technology, achieving 186 km altitude in 1946, launching seeds and fruit flies to study the effects of cosmic radiation a year later, making the first observation of the Sun's ultraviolet and X-ray spectrum and launching of primates to study the effects of acceleration and freefall on them in 1949.[42] The detonation of the Soviets' first nuclear device in September 1949 and the outbreak of the Korean War in 1950 forced the Americans to relook at their bomber-based strategy and turn it to place more reliance on missiles.

With the arms race well underway, von Braun and his rocket team in the United States developed the Redstone rocket 1952 onwards, the rocket that was to become the foundation for all future US missiles especially the Jupiter that was first launched in 1957. By the Jupiter's third launch, the rocket achieved a range of 1,600 miles, making it the first American Intermediate-Range Ballistic Missile (IRBM).[43] The desire to develop the ICBM manifested by assigning the rocket programme to multiple agencies, essentially the US armed Services. The US Air Force rocket programme evolved into the Atlas launch vehicle which launched the first American ICBM in June 1957, while von Braun[44] worked on the US Army's programme with the Jupiter launch vehicle.[45] The US Navy commenced its rocket programmes centred on three missiles: the Aerobee and the Viking sounding rockets, and the Submarine Launched Ballistic Missile (SLBM), the Polaris.[46] These starting blocks were then to go on to a whole range of rockets and missiles for both military and civil activities.

On the other hand, at the start of the arms race, the Soviets were clearly behind the Americans in terms of nuclear technology and the strategic depth of the air force and chose the decision to develop very large, high-thrust rockets for ballistic missiles. This reliance on heavy rockets was to confer on the Soviets a distinct lead in the early years of their forays into space. With the consensus arrived at in 1947 to build an ICBM, the Soviets refined their rocket designs through a series of R rockets. The R-7 Semyorka was notable amongst them, in that the huge missile built in 1955, was designed around a cluster of 20 rocket-motors and four boosters that were jettisoned in flight,[47] with twice the power of the contemporary American Atlas[48] and Titan ICBMs. The Soviets developed a whole range of missiles in the class of short- and medium-range intercontinental ballistic missiles. The SS-3 Shyster missile in 1956 was the world's first operational nuclear-tipped Medium

Range Ballistic Missile (MRBM)[49] while the SS-6 Sapwood powered by the ubiquitous R-7 rocket, became the world's first operational ICBM in August 1957.[50]

With the Cold War as a driver for the ideological and technological competition between the United States and the Soviet Union, the formative steps towards raising the envelope and exploring and exploiting space had commenced. The two superpowers would compete for the many 'firsts' that were to follow in the race for space.

Emergence of the Space Age

> *It is necessary also to master and release into the air other types of rockets as soon as possible in order to thoroughly study and attain adequate mastery of reactive techniques. Soviet rockets must reach space.*
>
> — Sergei Korolov

History changed on October 4, 1957, when the Soviets successfully launched the Sputnik I, the world's first artificial satellite into a Low Earth Orbit (LEO). The launch of the satellite, a little less than 2 ft in diameter and weighing just over 80 kg, was a singular ground-breaking event that, firstly, shattered the Americans' self-belief of technological superiority, and, secondly, in the larger perspective, ushered in the space age and the start of the great space race between the two superpowers, laying out the blueprint of the contest mentality in new political, military, technological and scientific developments, which would eventually shape the world into the way it is today.

While the run-up to the Sputnik I witnessed frenzied activity in matters of missiles and nuclear weapons, resulting in weaponry aboard rockets, scientists, visionaries, and collaborative military-civilian studies kept alive the dream of space flight. Project RAND, under the Douglas Aircraft Company,[51] was one such military sponsored think-tank (later becoming the RAND Corporation), which in its prescient report on "Preliminary Design of an Experimental World-Circling Spaceship" in May 1946, theorised the technicality of an artificial satellite in Earth orbit and spawned numerous subsequent reports on the feasibility of satellite design, launch, and reentry.[52] L. R. Shepherd and A.

V. Cleaver, in September 1948, published work on the technical feasibility of using nuclear power for space travel, while a British technical thinker, Arthur C. Clarke, two years later, suggested mining of the Moon and launching the mined lunar material into space with an electromagnetic catapult.[53] Millions of Americans were introduced to the concept of a space station and a human mission to the Moon in an article published in *Collier's* magazine in 1952. It was in the same year that von Braun published *The Mars Project*, the first serious technical study of a human expedition to Mars.

It was the same year, 1952, that the International Council of Scientific Unions designated the year July 1957-December 1958 as the International Geophysical Year (IGY), to explore the Earth and its atmosphere. This period was significant as scientists knew that the cycles of solar activity would be at a high point then.[54] In October 1954, the council adopted a resolution calling for artificial satellites to be launched during the IGY to map the Earth's surface. The United States, on July 29, 1955, publicly announced the intention to launch small Earth orbiting satellites as its involvement for the IGY, which four days later, was followed by the Soviets announcing a similar intent. On August 30, 1955, Sergei Korolov managed to create a commission whose sole purpose was to beat the Americans into space — this was the *de facto* date of the start of the space race.[55]

With apprehensions of using military rockets for launching satellites under the peaceful IGY programme and much against the advice of von Braun, the US President took the political decision of using the Vanguard rocket which was a research-only booster with the US Navy. This was despite the fact that by September 1956, US military rockets like the Redstone, Jupiter and Thor were ready to be launched.[56] Thus, the expertise that the US military possessed, especially in terms of funding, scientists and hardware to get the IGY mission accomplished, was not availed of for a larger and seemingly nobler cause. After several setbacks to the Vanguard project and smarting under the Soviet's Sputnik event of October 1957, the US President, in December 1957, approved a satellite launch using a modified Jupiter rocket. The distinction between military and peaceful rockets was, thus, blurred and von Braun was back in the fray. The first American satellite, Explorer 1, was finally launched on January 31, 1958, aboard von Braun's

Jupiter rocket[57] and its hallmark contribution to the IGY was the confirmation of the existence of the Earth-encompassing Van Allen radiation belt. Having learnt its lesson, the United States stuck thereafter to using modified military rockets for its space exploration right up to the 1980s.[58]

By the time the Americans achieved this national milestone, the Soviets had achieved yet another of the world's 'firsts'. Driven by Sputnik 1's success, Nikita Khrushchev demanded additional high-visibility space achievements from Korolov.[59] Laika, a female dog, the first living space traveller was launched into space aboard the Sputnik 2 on November 3, 1957. The Soviets took to pursuing exploration of the Moon too. It was the Soviets again whose spacecraft Luna 1 was the first to escape the Earth orbit en route to the Moon in January 1959. It was Luna 3 in October 1959 that beamed the first photographs of the Moon's other side.[60] Having an upper hand in rocket boosters, Korolov, egged on by Khrushchev, pressed home the Soviets' successes and launched the first human in space: Yuri Gagarin on April 12, 1961, orbited the Earth in a spacecraft named Vostok and was followed just three weeks later by America's Alan B. Shepard, Jr, launched into a sub-orbital mission, achieving the American 'first' on May 5, 1961. While Gagarin's Vostok flew on an automated mode, as the Soviets' medical science at that time was unsure of the effect of weightlessness on a human, Alan Shepard exercised manual control over his spacecraft.[61]

In the period 1958-61, both superpowers suffered a series of setbacks and also quite a few successes in satellite launches. In this four-year period the US had placed 48 satellites, including 41 for military purposes, into orbit, as against only 13 by the Soviets.[62] It was, however, a series of 'firsts' in space by the Soviets that continued to rancour the US, propelling them to realise that for the sake of international prestige, a major boost[63] would be required to close the gap with the Soviets. Speaking to the US Congress and the nation, US President John F. Kennedy, on May 25, 1961, boldly committed the United States to landing a man on the Moon by the end of the decade.[64] This clarion call galvanised the US National Aeronautics and Space Administration (NASA) into launching a whole series of Apollo missions. It was Apollo 11, launched aboard von Braun's Saturn V rocket[65] on July 16, 1969, that placed Neil A. Armstrong on the lunar surface four days later. The first step by a human on the Moon was witnessed by over 500 million TV viewers worldwide. *The Eagle had landed*, bringing glory back to its nation by the technological triumph of one of the biggest 'firsts' in the history of mankind.[66]

The early 1970s saw several more American lunar landings, more and more launches of technologically refined and versatile satellites from both the superpowers, probes to other planets of the solar system and deployment of space stations in LEO. This decade was also unique in that, contrary to the unwritten tenets of the Cold War, the decade ushered in the first tentative steps in the area of cooperation and collaboration between the two superpowers. The year 1975 witnessed a joint US-Soviet orbital rendezvous by manned spacecraft and by the 1990s, numerous countries worked together on space stations and on launching scientific missions.

Satellites: The Eyes in Space

Early successes in exploration of space led to frenzied activity in pursuance of national security and governmental policies and objectives towards preserving global stability during the Cold War. Satellites proved to be very effective tools in space, the medium that now came to be seen as the new 'high ground'. This section offers an insight into the progress made in the field of civilian and commercial satellite applications, that of communications, meteorology, Earth observation and navigation, while military applications of satellites are covered in detail in Chapter 7.

Communications Satellites: Communications satellites are one of the most significant types of satellites enabling people to communicate by telephone, exchange data over vast distances and relay television and radio signals to the ground for entertainment and educational purposes. Interestingly, the first actual communications satellite was the Moon itself. In the 1950s, the US Navy developed a system for bouncing signals off the surface of the Moon to communicate to and fro between the continental United States and Hawaii. This system was crude and worked only when the Moon was in the right position, but it demonstrated the capability that objects in orbit around the Earth could aid in communicating over long distances.[67] As early as in 1945, British scientist Arthur C. Clarke proposed placing three satellites evenly spaced around the globe in a geosynchronous orbit that could view the Earth's entire surface and, therefore, be able to relay communications for the entire planet.

> *This is the President of the United States speaking. Through the marvels of scientific advance, my voice is coming to you from a satellite traveling in outer space. My message is a simple one: through this unique means, I convey to you and all mankind, America's wish for peace on Earth and goodwill toward men everywhere.*
>
> — US President Dwight D Eisenhower, 1958

This was the US President's Christmas message that was beamed from space under Project Score and was unique, in that, it was the world's first transmission from a satellite. In those days, telephone communications took place across the oceans through a huge network of undersea copper cables. These cables were expensive to put in place and maintain, and satellites represented a potentially cheaper way to accomplish the same task. In particular, the US communications giant American Telephone and Telegraph (AT&T) wanted to develop them but certain quarters in the US government had apprehensions that AT&T could gain a monopoly in this new field. Despite this apprehension, US President John F. Kennedy signed the Communications Satellite Act of 1962 that created the Communications Satellite Corporation.[68] A direct outcome of this Act was in July 1962 when the first commercially funded satellite, Telstar, funded by AT&T, was placed in a LEO, providing communication access to the US, United Kingdom and France. To obviate the technicality involved in tracking the satellite from the Earth and to simplify the associated process of antenna alignment and pointing, satellites in Geostationary Earth Orbit (GEO) were felt to be a necessity. The first communication satellite to operate from GEO was NASA's Syncom, which also broadcast the Tokyo Olympic Games in 1964, live over television from Japan to the US. The age of instantaneous global communications had arrived. The era of international cooperation and collaboration was heralded in 1964 when telecommunications agencies of 18 nations formed the International Telecommunications Satellite (INTELSAT)

Organisation, to coordinate satellite communications in a manner of sharing the cost of satellite communications based on the amount of usage. A year later, INTELSAT launched its first geosynchronous communications satellite, INTELSAT 1, the Early Bird, which provided almost 10 times the capacity of the submarine telephone cables at almost one-tenth the cost.[69] This satellite was the first of many and the service achieved global coverage by 1969. By the late 1970s, the INTELSATs were serving close to 100 nations and due to the increasing demand, appropriately, the capacity and weight of the satellites grew, and continues to do so till date.

In the Soviet Union, communications satellites came of age with the launch of its Molniya 1A satellite in April 1965,[70] in a highly elliptical 12-hour orbit. Many other countries, despite being participants of the INTELSAT initiative, driven by domestic compulsions developed indigenous communications satellites. Canada was the first, launching its Anik satellite in 1972. Others followed: Indonesia's Palapa, India's INSAT, Australia's Aussat, Mexico's Morelos, and Brazil's Brazilsat. In addition, 22 Islamic nations formed a regional group, the Arab Space Communication Organisation, launching its first satellite, the Arabsat, in 1985.[71] As more and more communications satellites began being placed in GEO, satellite designers and operators were presented with the problem of their satellite signals interfering with each other. This required careful international regulation and meant that satellites were assigned specific spots. Jostling for GEO "real estate", thus, became an important issue for operators and was one reason for the development of ever larger satellites, able to pack more transmission capability into a single satellite.

Commencing in the late 1980s, satellite orbits for communications underwent a relook as several companies began considering placing satellites in LEO, employing hand-held receivers to communicate with them. The first such system was operated by a US consortium named Iridium, which went into operation in the late 1990s. Other companies followed suit but their systems required many satellites with sophisticated equipment for relaying signals from one satellite to another.[72] As of today, communications satellites see tremendous employment in covering a very wide spectrum of applications and services, ranging from telephony, including cellular services, television, video-conferencing, Internet, distance learning and education, and radio communications, including amateur radio.

Satellites for Meteorology: Weather satellites have a unique advantage of observing weather patterns that are obscured from the ground. Meteorologists even before the space age had realised the immense value of weather monitoring instruments at a sufficient height above the Earth that would enable them to gather data to provide accurate weather forecasts. The arrival of the age of satellites provided the impetus towards realising the weatherman's dream. The world's first weather satellite, TIROS 1, capable of imaging clouds from a polar orbit was launched by the US' NASA in April 1960. This was the first step in satellite-based meteorology that opened the door to deeper understanding of the natural forces that affected terrestrial weather. The family of TIROS satellites was followed by the Nimbus in 1964, a new series of advanced weather satellites designed to meet the needs of the world's atmospheric science research community. Orbiting in low altitude near-polar Sun-synchronous regions, this system provided mosaic images of the entire planet. During this period, US military weather satellites under its Defence Meteorological Satellite Programme (DMSP) came into being, capable of providing joint force commanders with essential data required for accurate, dependable weather forecasting in support of air, land, and maritime operations, and eventually playing a significant role in national weather forecasting for defence and civil activities.[73] GEO orbits began being occupied by weather satellites when NASA launched the Geostationary Operational Satellites series of satellites. Meanwhile in Europe, the European meteorological satellite system, Meteosat, a collaborative effort of eight member nations of the European Space Agency (ESA), launched Meteosat-1 in November 1977 from Cape Canaveral, Florida, and the balance of a series of GEO Meteosats from French Guyana. The advantages accrued by weather satellites in GEO is that equipped with a vast array of onboard sensors, imagers and instruments, these satellites provide continuous monitoring over the full-disk view of the Earth, providing invaluable inputs towards forecasting and disaster management resulting from severe weather conditions such as hurricanes, tropical storms and flash floods. Thus, while GEO weather satellites provide short-term and 'now-casting', weather satellites in polar orbits provide longer-term forecasting and data from both the input sources needs to be blended and analysed to arrive at a comprehensive global weather picture.

Satellites for Earth Observation: As an offshoot from the technology driven quest for weather mapping, another big step was taken in 1972 when special payloads aboard satellites began monitoring the Earth's land and sea surfaces in great spatial and spectral detail. The US launched its Earth Resources Technology Satellite-1 (ERTS-1) in July 1972, into a Sun-synchronous polar orbit. As part of the Landsat series of satellites, the scientific community was now enabled to study geology and natural resources in all corners of the planet. The age of remote-sensing had arrived with extensive capability in the fields of environmental monitoring, map making, assessing vegetation growth, monitoring the spread of cities, and collecting and analysing data on how the Earth's surface was changing. In the Soviet mainland, similar activity commenced in the late 1980s with the Okean series of satellites, which produced radar maps of the Earth's surface and were also used to measure waves on the oceans' surface. The USA took the lead in multi-dimensional simultaneity, allowing scientists to view the Earth as a dynamic and integrated system. The Terra series of polar orbiting satellites, the first of which was launched in December 1999, look at the Earth's radiation balance, atmospheric circulation, biological productivity, air-sea interactions and natural hazards identification.[74]

The dynamics of remote sensing are such that they spawn intense interest amongst multiple agencies like those for urban planning, terrain mapping, oceanography, water resource management, minerals, agriculture and forestry. To capitalise on such benefits, other nations have travelled the indigenous route and have built their own operational remote-sensing satellites. India for one, has not only built its own remote-sensing satellites but has placed them in orbit using its own launch vehicles. Commencing from rather humble beginnings when it launched its first remote sensing satellite, the IRS-1A, aboard a Soviet rocket in March 1988, and continuing since 1994 with its own launch capability, the Polar Satellite Launch Vehicle (PSLV), the Indian Remote Sensing Satellite system today fields the world's largest operational constellation of remote sensing satellites for civilian use. Europe's contribution commenced with the European Space Agency's (ESA's) Europe Remote Sensing Satellite, the IRS-1, launched in 1991. Canada's Radarsat, launched in 1995, forms its images by using radar instead of visible light. The Argentinian spacecraft, the Scientific Applications Satellite or SAC-C, launched in November 2000, carries remote-sensing equipment along with other payload. In Asia, besides India, South Korea pursued its remote sensing programme with the launch of two 100-pound satellites in 1992 and 1993. The programme, called Uribyol or Our Star, has doubled its capability with its third satellite that was launched in 1999. China was rather late in entering the field of remote sensing, with the first launch of its Ziyuan satellites in 1999 aboard the Long March rocket.

Satellites that Aid Navigation: Amongst the four classes of satellites considered in this section, there are two that impact the common man directly, across the world. The two that bring satellite-enabled services and applications right into a man's home are the communications and navigation satellites. What is common to both these classes of satellites, in addition, is their genesis. The *raison d'etre* for the development of both comprised matters military – they were purely for military applications, with commercial services and applications emerging subsequently as natural offshoots. The era of space-based navigation commenced in April 1960 with the launch of the US Navy's first experimental navigation satellite, the Transit 1. The satellite relied on the Doppler shift of frequencies in the satellite's radio signal due to the shift in the position of the satellite, which enabled ships and submarines to accurately fix their positions. The fourth in this series, the Transit 4A, was also the world's first nuclear powered satellite. This system had a large margin of error and was unsuitable for application in the aviation sector, for which the solution lay in the Global Positioning System (GPS). The NAVSTAR GPS family of US satellites, with the USAF as its designated executive Service, finally expanded to a constellation of 20 operational satellites by March 1994, providing an accuracy of a few metres.[75] The technical requirement would be that of a 24 GPS satellite constellation, in semi-synchronous orbit (12-hour orbit at an altitude of 20,200 km) in six orbital planes, ensuring that at least four satellites are in view by the user at all times.[76] This GPS provides for an accuracy of 2-4 m for military users and 3-5 m for civilian customers worldwide.

The value-addition that GPS provides to the potency of any nation's military services, as well as the tremendous scope that accrues to the civil and commercial realms of a nation's geopolity, as good as mandates a forward-looking nation to address indigenous efforts in harnessing the potential of GPS and achieving self-sufficiency in this technology driven field. With the USA's GPS achieving the global status of the most utilised satellite navigation system, the Soviet Union's Global Navigation Satellite System (GLONASS) became fully operational in 1995, but fell into partial availability after the collapse of the Soviet Union, and then recovered, and global coverage was restored in 2011. The Chinese Beidou navigation system, operational since 2000, based on over a dozen satellites is a regional system that provides coverage to China and neighbouring regions, providing an accuracy of 10 m with plans of providing global coverage by 2020.[77] The European Union (EU), with ESA, launched the first of its navigation satellites in 2005 under its Galileo navigation system which is expected to turn global by the end of this decade. India's foray in this field revolves around ISRO's Indian Regional Navigational Satellite System (IRNSS), based on seven satellites and planned to be in place by 2014[78] providing regional coverage to areas covering 1,500 km around the country.[79]

GPS receivers today are being developed for both military and civil-end users, and are commonly used in surface vehicles, aircraft and ships, with commercial hand-held GPS receivers commonly available in use with cell phones, computers and cars.

Laboratories in Space

By the late 1960s, the Soviets had more or less come to terms with the fact that the Americans were going to win the race for the 'man on the Moon'. As a consequence, they redirected their focus towards long duration missions in LEO, by relying on orbiting space platforms that could sustain relatively long duration human habitation. The Salyut 1, the world's first space station, was a product of this vision. Launched on April 19, 1971, it functioned in a civil research capacity as an observation platform gathering data on the Earth's resources, meteorology and astronomy.[80] Three cosmonauts aboard Soyuz 11 docked with the Salyut 1 three months later and spent 24 days on the station. Tragedy occurred when the three cosmonauts were killed during their reentry into the Earth's atmosphere. Future flights were cancelled and the Salyut 1 was taken out of orbit after only 175 days.[81] The Salyut 3, the first operational 'military mission' space station, launched in December 1974, had the crew conduct military photoreconnaissance operations of which not much is known. The Salyut series[82] of Soviet space stations evolved into their third generation space station, Mir, which was assembled in orbit around a core module launched in February 1986. This massive space station, built with six docking ports, was able to receive supply 'ships' and was visited not only by cosmonauts, but also by guest researchers, including American astronauts aboard their Space Shuttle. Mir was finally abandoned and safely deorbited in March 2001, due to economic constraints and the collaborative work that had commenced on the International Space Station (ISS) in 1998.

The Americans' endeavour towards their own space station had actually commenced in 1963,[83] well before the lunar landing. They envisioned a two-pronged strategy in fulfilment of their next giant step in crewed space flight. Firstly, a safe and reliable transportation system to enable routine access to space, and, secondly, an orbiting space platform where humans could live and work akin to a base camp from where other advanced space technology programmes could be initiated. This paved the way for NASA's Skylab, launched in one singular piece in May 1973 aboard the Saturn rocket (of the Apollo fame), which was designed for a variety of experiments such as the effects of long-term weightlessness and zero gravity. The Skylab was not designed as a permanently orbiting platform, and in July 1979, met its fiery end over the Indian Ocean and Western Australia.[84] As a result of US President Ronald Reagan's initiative to bring in US-allied partners for the space station programme, Japan, Canada and the European Space Agency (ESA) joined it in 1985. By 1992, with the Cold War as good as over, the Russian collaborative work too provided a significant contribution to the ISS. In December 1993, the United States agreed to pay US$ 472 million to Russia, to permit American astronauts to visit Mir from 1999 onwards and for Russia to provide a propulsion module for the ISS.[85] The ISS, the largest space station ever constructed, is also one of the most ambitious international collaborations ever attempted.[86] Spread across three phases of development commencing in 1998, the ISS was literally constructed and deployed in space, module-by-module by the American Space Shuttles and the Russian Soyuz and Proton rockets. The ISS programme is a joint project between five participating space agencies, the American NASA, the Russian Federal Space Agency (RKA), the Japan Aerospace Exploration Agency (JAXA), the European ESA, and the Canadian Space Agency (CSA)[87] and has been visited by space travellers from 15 different nations – and counting – actively involving themselves with study and research in the primary fields of astrobiology, astronomy, human research, including space medicine and life sciences, physical sciences, material science, space weather and meteorology. When finally assembled, it will span a mammoth 110 m across and 90 m in length, with almost an acre of solar panels to provide power to six state-of-the-art space laboratories providing an observational coverage of 85 percent of the globe.[88] It continues to receive and rotate crew and space tourists from the Russian Soyuz manned spacecraft and is resupplied by the Russian Progress cargo spacecraft, Japan's H 2 Transport Vehicle (HTV) and ESA's Automated Transport Vehicle (ATV). Since May 2012, the ISS is also being resupplied by the world's first privately owned commercial space transportation company, the Space Exploration Technologies Corporation, or Space X, through their Dragon spacecraft launched aboard the company's Falcon launch vehicle.

Reusable Space Vehicles

While the largely successful conduct of space travel by humans showed that the era of safe space travel had arrived, rockets were still the only propelling source relied upon by the spacefaring nations. These rockets were large and costly, being limited to single use and discarded thereafter. It was in the 1970s that NASA began considering the concept of 'reusable' launch systems. US President Richard Nixon in 1972, gave the go-ahead for the Space Shuttle, an inexpensive manned 'space truck' that, as a reusable transportation means would move humans and payloads quickly in and out of space, a policy decision that would shape a major portion of NASA's programme for the next three decades,[89] as the Space Transportation System (STS). Based around the core of a reusable three-engined, double delta-winged orbiter craft, the STS also had a giant external expendable tank, and, in addition, two jettisonable but reusable solid-rocket boosters. The STS was launched vertically upwards akin to any other space launcher but its uniqueness lay in its ability to land on a runway like a conventional aircraft. Almost a decade of development later, NASA launched its first STS on April 12, 1981. In all, six Space Shuttles were built and employed across a wide spectrum of space activities spread over three decades.

Apprehensions regarding the Americans using the Space Shuttle for undeclared military purposes made the Soviets develop their own orbiter programme. The Buran programme saw only one unmanned space flight in 1988 while four others underwent only a few trial flights. The Soviet Buran programme was finally cancelled in 1993.

The shuttle proved to be a valuable system for understanding the challenges of living and conducting scientific experiments in space. Transporting astronauts to Mir and the ISS, repair of the Hubble telescope, transporting the Spacelab (ESA's self-contained scientific and engineering laboratory), deploying satellites in their orbits, assisting in repair and construction of the ISS, deploying NASA's Magellan spacecraft on its voyage to Venus, and recovery of satellites, were some of the more significant tasks carried out by the Space Shuttles. Despite these successes and the technological superiority that the US enjoyed as the only nation with a successful reusable STS, the programme was found wanting in its primary goal, that of being an inexpensive 'space truck'. The fast turnaround that was envisaged as a key to the economics of space travel was limited by the heavy maintenance requirements of the Space Shuttle. Initial costs and expensive upkeep made it prohibitively expensive to launch satellites.[90] The Shuttle was enormously expensive to fly and every flight cost between $400 million and 1 billion.[91] The US STS programme was, thus, officially terminated and finally ended when the last flight of Space Shuttle Atlantis landed at the Kennedy Space Centre on July 21, 2011. In the preceding three decades, the Space Shuttles flew 135 missions,[92] including 37 for servicing the ISS.

The sense of adventure, has, from times immemorial, led man to seek newer and newer horizons, the path to which led to discoveries at times, while, at other times, man had necessarily to innovate to achieve his goals.

At no point in history has the human race had such a mixed bag of fortune on his path to progress as was witnessed in his quest for taking to the skies and then beyond – to space. The last century took man from the Earth to the clouds above him and then onwards to newer horizons in space. The alluring mysteries of the skies above beckoned the likes of the Wright Brothers while the heavens beyond triggered the minds of Tsiolkovsky, Goddard and Oberth.

Much of the initial innovative work in space exploration owed its genesis to the service of the militaries, in pursuit of innovation-led technology drivers to achieve national military goals and objectives. The end of World War II gave way to the Cold War – rockets of the Great War evolved into missiles of the post-war and attempts to leverage the potential of space resulted in spy satellites and weapons in space. The race for man-on-Moon came about as a result of the race for technological superiority between the two competing superpowers. The year 1991, the end of the Cold War, may well be termed as the year when commercial and collaborative concerns overrode the military pursuits in matters of exploiting space. Striving for peaceful pursuits in space came to be the new *mantra*, which now sees nations across the world working together and also independently in carving out for their citizens a piece of the space-pie, working towards innovation and exploration in making the *global village* a more comfortable and habitable planet to live in.

The Cold War was an event of the past. The break-up of the Soviet Union in 1991 has changed the global geopolitical landscape. The benefits that accrue from pursuits in space are now sought by over 60 countries that have assets, in some form or the other, in space. The subsequent chapter will cover some of the significant space players and their standing in the emerging global order.

Notes

1. The two-man crew of this manned balloon could see details on the ground up to 29 km away with the use of a telescope. Scientist Charles Coutelle, assisted by N.J. Conté, demonstrated their innovation to members of the French Commission, who were impressed enough that they recommended formation of an air force, the world's first, called the *Compagnie d'AΘronautiers*, which was finally established on March 29, 1794, three months before the Battle of Fleurus in June 1794. http://www.centennialofflight.gov/essay/Lighter_than_air/Napoleon's_wars/LTA3.htm,

accessed on January 30, 2012.

2. Somnath Sapru, *Armed Pegasus: The Early Years* (New Delhi: KW Publishers Pvt Ltd, 2010), Ch. 1, p. 22.
3. Joseph A. Angelo, Jr, *Space Technology* (New Delhi: Pentagon Press, 2009), p. 6.
4. Ibid.
5. http://www.solarviews.com/eng/rocket.htm, accessed on February 1, 2012.
6. Angelo, n. 3, p. 6.
7. http://space.au.af.mil/au-18-2009/au-18_chap01.pdf, accessed on February 1, 2012.
8. Wg Cdr Kaza Lalitendra, *Militarisation of Space* (New Delhi: KW Publishers Pvt Ltd, 2010), p. 2.
9. One of Tipu Sultan's rockets continues to be displayed in the Royal Ordnance Museum at Woolwich Arsenal, near London.
10. http://www.solarviews.com/eng/rocket.htm, accessed on February 1, 2012.
11. Joseph A. Angelo, Jr, *Encyclopedia of Space and Astronomy* (New Delhi: Viva Books Private Limited, 2010), p. 554.
12. Sqn Ldr K.K. Nair, *Space: The Frontiers of Modern Defence* (New Delhi: Knowledge World, 2006), p. 2.
13. Marianne J. Dyson, *Space and Astronomy: Decade by Decade* (New Delhi: Viva Books, 2010), Ch. 1, p. 10.
14. Konstantin Tsiolkovsky's tombstone in the old Kaluga cemetery, not far from the museum that honours his life and work, bears the prophetic inscription: "Mankind will not remain tied to the Earth forever".
15. Angelo, n. 11, p. 515.
16. Although Tsiolkovsky was less well known during his lifetime in the United States, his work enjoyed broad study in the 1950s and 1960s as Americans sought to understand how the Soviet Union had accomplished such unexpected successes in its early efforts in space flight. American space scientists then realised that his theoretical efforts had been essential for the development of the practical rocketry on which the Soviet space programme was based. http://www.centennialofflight.gov/essay/SPACEFLIGHT/tsiolkovskiy/SP1.htm, accessed on February 14, 2012.
17. Angelo, n. 3, p. 9.
18. Angelo, n. 11, p. 515.
19. Goddard's work remained largely unacknowledged by the government and the US military during his lifetime, except for government funding during World War II for work on rockets to assist take-off of naval carrier-borne aircraft. Help came his way by a $ 50,000 grant through the famous aviator Charles Lindbergh which enabled Goddard to shift to a remote part of New Mexico for his subsequent experimentation in the field of rocketry.
20. It was five years later that the first German liquid-fuelled rocket was launched, on

March 14, 1931, while the first Russian liquid-fuelled rocket, the GIRD-09, was launched on August 17, 1933. Dyson, n. 13, Ch 4, p 79.

21. Angelo, n. 11, p. 427.
22. Lalitendra, n. 8, p. 3.
23. Dyson, n. 13, Ch. 4, p. 82.
24. http://space.au.af.mil/au-18-2009/au-18_chap01.pdf, accessed on February 13, 2012.
25. Ironically, it was von Braun's rocket technology that enabled the United States to send a man to the Moon three decades later.
26. V-2 rockets were expensive and required numerous test flights to confirm changes in design. Metals were in short supply and so was ethyl alcohol that was used for fuel. A humorous anecdote is provided by Marianne J. Dyson in *Space and Astronomy: Decade by Decade*, of the workers pilfering the ethyl alcohol to make "instant schnapps" to drink. (One rocket used alcohol equivalent to about 66,ooo dry martinis). Adding of a purgative to the fuel was tried as a solution but the resulting absenteeism almost brought the test programme to a halt.
27. Dyson, n. 13, Ch. 4, p. 83.
28. Angelo, n. 3, p. 11.
29. http://www.aviation-central.com/space/usm10.htm, accessed on February 14, 2012.
30. Michael J. Neufeld, *Von Braun: Dreamer of Space, Engineer of War* (New York: Alfred A. Knopf, 2007), p. 184.
31. Lalitendra, n. 8, p. 22.
32. http://www.centennialofflight.gov/essay/SPACEFLIGHT/von_Braun/SP4.htm, accessed on February 14, 2012.
33. Operation Crossbow was the Allied campaign involving operations against all phases of the German long-range weapons programme—operations against research and development of the weapons, their manufacture, transportation, their launching sites, and against missiles in flight. The V-2's supersonic speed and high altitude trajectory made it invulnerable to the existing air defences of that time and Operation Crossbow, thus, largely concentrated on aerial bombing of the rockets' ground-based infrastructure. The campaign included Operation Hydra, the Royal Air Force attack on the Peenemünde Army Research Centre on the moonlit night of August 17/18, 1943, which, by most accounts, was not as successful as planned, though it did delay the V-2 test launch schedule by about seven weeks.
34. Dyson, n. 13, Ch. 5, p. 96.
35. Paperclip, the project's operational name, was derived from the paperclips used to attach the scientists' dossiers to the new political personae in the American personnel files.
36. Dyson, n. 13, Ch. 5, p. 97.

37. The US Army was later to have estimated that the American-German 'collaborative' research and development in this field, gained the US 10 years of research and saved the nation at least US$ 750 million. Dyson, n. 13, Ch. 5, p. 100.
38. Asif A. Siddiqi, *Challenge to Apollo: The Soviet Union and the Space Race, 1945-74* (Washington DC: NASA History Series, 2000), pp. 82-84.
39. William E. Burrows, *This New Ocean: The Story of the First Space Age* (New York: Random House, 1998), p. 147.
40. Dyson, n. 13, Ch. 5, pp. 98-99.
41. http://space.au.af.mil/au-18-2009/au-18_chap01.pdf, accessed on February 1, 2012.
42. Dyson, n. 13, Ch. 5, p. 100.
43. Erik Bergaust, *Rocket City, U.S.A.: From Huntsville, Alabama to the Moon* (New York: Macmillan, 1963), p. 90.
44. It would be the Redstone rocket in its modified versions that would launch both America's first satellite and the first piloted Mercury space missions. Burrows, n.39, p. 137.
45. Dyson, n. 13, p. 122.
46. http://space.au.af.mil/au-18-2009/au-18_chap01.pdf, accessed on February 1, 2012.
47. Angelo, n. 3, p. 12, & Dyson, n. 13, p. 122.
48. The R-7 weighed 23 tons as against 7.7 tons of the US' Atlas.
49. Siddiqi, n. 38, pp. 120-21.
50. Rex Hall, David J. Shayler, *The Rocket Men: Vostok & Voskhod, The First Soviet Manned Spaceflights* (New York: Springer–Praxis Books, 2001), p. 56. Not only the first ICBM, it was the R-7 and its subsequent refinements that launched the world's first satellite (Sputnik 1), the first human into space, and the first lunar and planetary probes.
51. http://www.space4peace.org/slaw/lawofwar.htm, accessed on February 21, 2012.
52. Bert Chapman, *Space Warfare and Defence* (California: ABC-CLIO Inc, 2008), pp. 5-6.
53. Angelo, n. 3, p. 12.
54. http://history.nasa.gov/sputnik/, accessed on February 28, 2012.
55. James Schefter, *The Race: The Uncensored Story of How America Beat Russia to the Moon* (New York: Doubleday, 1999), pp. 3-5.
56. http://space.au.af.mil/au-18-2009/au-18_chap01.pdf, accessed on February 1, 2012.
57. Von Braun remarked on the event, "We have firmly established our foothold in space. We will never give it up again". Dyson, n. 13, p. 125.
58. Angelo, n. 3, p. 14.
59. Angelo, n. 11, p. 338.
60. Ibid., p. 339. One of the largest features on the far side of the Moon appropriately is named after Korolov. Besides these pioneering works, Korolov is also credited with

other 'firsts', like human in space (1961), spacecraft to Mars (1962), woman in space (1962) and the space-walk (1965). Korolov is, thus, rightfully regarded as the rocket engineer who started the space age.

61. Kenneth Gatland, *Manned Spacecraft, Second Edition* (New York: MacMillan Publishing Co, Inc, 1976), p. 101.
62. Lalitendra, n. 8, p. 8.
63. NASA's budget went from US$ 500 million in 1960 to 5.2 billion in 1965, half of which went into human space flight. Dyson, n. 13, p. 148.
64. http://history.nasa.gov/SP-350/ch-2-1.html, accessed on March 6,2012. He proclaimed, "First, I believe that this nation should commit itself to achieving the goal, before this decade is out, of landing a man on the Moon and returning him safely to the Earth. No single space project in this period will be more impressive to mankind, or more important for the long-range exploration of space; and none will be so difficult or expensive to accomplish." He justified that this endeavour was not only vital for national security but would also enable focussing on scientific and social fields. The public sentiment and national fervour was high and willing to expand the space effort, and the US Congress did not bat an eye at the projected estimate that $20 billion to $40 billion could be required to go to the Moon. It was a popular decision, and the vote of the Congress on the Moon programme was virtually unanimous.
65. One of the craters on the Moon is named after von Braun.
66. It was not only the single-minded and focussed zeal of the NASA team that led to this unique feat. As compared to the Americans, the Soviets just simply lacked the massive funding that the Americans pumped into the 'Man on Moon' mission. In the USSR, Khrushchev's removal from power in 1964 precipitated Korolov's falling from grace and the position of power that he enjoyed under Khrushchev's patronage. Korolov's death in 1966, at the age of 59, cut short the plans of the N-1 rocket which was being developed for the Soviets' manned Moon landing mission.
67. http://www.centennialofflight.gov/essay/SPACEFLIGHT/comsats/SP43.htm, accessed on March 16, 2012.
68. http://www.centennialofflight.gov/essay/SPACEFLIGHT/comsats/SP43.htm, accessed on March 16, 2012.
69. http://www.hq.nasa.gov/office/pao/History/satcomhistory.html, accessed on March 19, 2012.
70. Angelo, n. 3, p. 25.
71. http://www.centennialofflight.gov/essay/SPACEFLIGHT/intelsat/SP44.htm, accessed on March 19, 2012.
72. http://www.centennialofflight.gov/essay/SPACEFLIGHT/comsats/SP43.htm, accessed on March 19, 2012.

73. Angelo, n. 11, p. 397.
74. Angelo, n. 3, p. 28.
75. Ibid., p. 26.
76. http://space.au.af.mil/au-18-2009/au-18_chap16.pdf, accessed on March 21, 2012.
77. http://www.bbc.co.uk/news/technology-16337648, accessed on March 22, 2012.
78. http://www.isro.org/scripts/futureprogramme.aspx, accessed on March 22, *2012.*
79. http://www.thehindu.com/sci-tech/article393892.ece, accessed on March 22, 2012.
80. Angelo, n. 11, p. 524.
81. Angelo, n. 3, p. 18.
82. It was the Salyut 7 that hosted amongst others, the one and only Indian cosmonaut, Wg Cdr Rakesh Sharma.
83. Angelo, n. 11, p. 569.
84. Ibid., p. 569.
85. Dyson, n. 13, p. 220.
86. http://www.nasa.gov/mission_pages/station/structure/elements/partners.html, accessed on March 13,2012.
87. Gary Kitmacher, *Reference Guide to the International Space Station* (Canada: Apogee Books, 2006), pp. 71–80.
88. http://space.au.af.mil/au-18-2009/au-18_chap01.pdf, accessed on February 1, 2012.
89. Angelo, n. 3, p. 19.
90. Ibid., p. 20.
91. http://www.centennialofflight.gov/essay/SPACEFLIGHT/Shuttle/SP25.htm, accessed on March 13, 2012.
92. http://www.nasa.gov/mission_pages/shuttle/flyout/index.html, accessed on March 21, 2012.

3

Other Players in Space

Ships and sails proper for the heavenly air should be fashioned. Then there will also be people, who do not shrink from the dreary vastness of space.

— Johannes Kepler to Galileo Galilei, 1609

The Cold War that commenced between the two superpowers at the end of World War II, as a natural follow through, saw the rivalry reaching space that led to an intense desire to achieve the many *firsts* in space. The end of this rivalry in 1991 witnessed budget cuts in both the players that brought about a slackening in the pace of space related development. By this time, other nations too, seeing the tremendous potential of this medium and driven by the domestic compulsions of sovereignty, had commenced their own exploratory forays into space. Countries across the globe scrambled to get on the space-bandwagon, some preferring the collaborative and cooperative approach among themselves while others went the independent route. Building satellites is technologically challenging and yet has been achieved by over 60 countries worldwide – what is more challenging is the means of launching and putting them in orbit. The latter has been achieved, other than by the USSR/Russia/Ukraine and USA, by only nine other countries. The two superpowers could well be said to have ushered in the 'first wave' of spacefarers, which were followed by the 'second wave' of launch-capable countries, in chronological order: France (1965), Japan (February 1970), China (April 1970), Britain (1971), India (1980), Israel (1988), Iran (2009), North Korea (December 2012) and South Korea (January 2013). In so doing, they achieved the capability to build their own satellites and then to launch and place them in orbit, while a few went a few steps ahead and achieved the status of emerging global space powers. This chapter will trace out the historical profiling of the 'second wave' of spacefaring entrants in the emerging global field of space players and briefly look at some of the prominent members of the 'third wave' in and around the Indian subcontinent, countries with a fairly well established satellite technology base but not yet having achieved a space-launch capability.

European Space Agency (ESA)

The Genesis

Post World War II, while the US and the Soviet Union capitalised on taking off from von Braun's expertise in rocketry, most of Europe occupied itself in post war reconstruction. It was the launch of the Sputnik 1 in October 1957 that prompted members of the scientific community in eight countries of Western Europe to meet a few months later, in 1958, to address the need of founding a common Western European Space Agency (ESA). The decision to take the collaborative route by pooling efforts and resources was essentially economic, as no individual European country on its own could afford to support the complete range of space activities and the requisite infrastructure (though some countries, such as France,[1] do have a substantial national space programme of their own, in addition to being a member of the ESA).[2] It was decided to establish two agencies, one for developing launch systems and the other for space research systems, the European Space Research Organisation (ESRO) which was eventually established in March 1964. ESRO 2B, the first European research satellite, launched on May 17, 1968, was followed by six more in the next four years, all aboard NASA's launchers. ESA, as it is known today, emerged in 1975 on the merging of the launch and research establishments and bringing both within the framework of ESA. ESA's first major scientific success was in that year itself with the launch of the Cos-B, a radiation-monitoring space probe.

Arianespace, ESA's arm that since 1984 commercially markets the Ariane launch vehicles, commenced its operations from Kourou, French Guyana, in South America, in 1970, and launched the first Ariane rocket in December 1979. It was Ariane-1 that launched the Giotto spacecraft in July 1985 from French Guyana, on a mission to encounter and investigate Halley's Comet, which it did in March 1986. By the late 1970s, confident of its launch potential from French Guyana, ESA had already entered into a partnership with the US' NASA in numerous projects in the field of space research. The world's first high-orbit telescope was launched in January 1978 aboard the International Ultraviolet Explorer, jointly operated by NASA and ESA. Spacelab, an orbiting space laboratory developed by ESA in cooperation with NASA, was placed in orbit aboard the Space Shuttle in November 1983 and subsequent missions saw ESA astronauts participate in experiments on the laboratory. This was followed by the Hubble space telescope launched aboard the Space Shuttle into a LEO in April 1990, again in cooperation with NASA. Hubble which is still functioning as of date, is a space-based optical observatory for the benefit of the international astronomical community to observe the visible universe. The Ulysses spacecraft was built by Germany's Dornier Systems for ESA and launched by NASA's Space Shuttle in October 1990, with an aim to study the unexplored region of space above the Sun's poles. The Solar and Heliospheric Observatory (SOHO) was yet another spacecraft designed for study of the Sun which was launched by the US' Lockheed Martin Atlas launch vehicle in December 1995.

The Age of Cooperation

The saga of international cooperation in space can best be exemplified by the International Space Station (ISS). Originally conceived by the US, the ISS called for the participative involvement of US-allied countries. Japan, Canada and ESA signed bilateral agreements with the US in 1985 with Russia joining in 1992. ESA's involvement in the ISS since then has been the Automated Transfer Vehicle (ATV) which supplies logistics to the ISS, and Columbus, a pressurised science laboratory module as an add-on to the ISS.

Today, ESA has broadened the scope of its participative collaboration and has entered into cooperation agreements with Russia, China, India, Japan, Argentina and Brazil. ESA's spectrum of space activities spans the fields of space sciences, Earth observation, telecommunications, space segment technologies (including orbital stations and platforms), ground infrastructure, space transportation systems, and basic research in microgravity,[3] while also maintaining its own corps of astronauts. In the Ariane-5 space launch vehicle, the space-workhorse for the European community, ESA sees the capability of placing a wide variety of payloads in LEO, GEO and for interplanetary missions, for both inhouse and commercial ventures. Within ESA's launcher family, the Ariane has been joined since 2011 by the Russian space-horse, the Soyuz rocket and in 2012, by the Italian Vega rocket that places smaller payloads into a 700 km orbit. Ariane-6, to be fielded in 2021-22, aims to garner the lucrative commercial launch segment by cutting down the launch costs of the current space-horse, the Ariane-5, by 30 per cent.

Military Applications

ESA's stated purpose is to provide for, and promote, for exclusively peaceful purposes, cooperation among the European states in space research and technology and their space applications.[4] However, ventures have been made in the field of military applications too. The military space capability, though not ESA-initiated, has been largely formed by national programmes or inter-governmental collaborations.[5] The UK has the longest running national military satellite programme in Europe, including satellites for electronic signals intelligence and surveillance. Other European countries with significant engagements in military space are France, Germany, Italy and Spain, while some of the smaller countries have limited military space programmes, often sharing the facilities and infrastructure of the main players.

ESA and the European Union

While ESA is not a body or an agency of the European Union (EU), the two remain separate organisations which are increasingly working together towards common objectives. In recent years, the ties between the two institutions have been reinforced by the increasing role that space plays in supporting Europe's social, political and economic policies leading to 20 per cent of the funds managed by ESA now originating from the EU budget. With a yearly budget of over US$ 5 billion,[6] member countries of ESA engage in extensive cooperation and implement a wide range of significant space projects for scientific and commercial purposes. In the years to come, ESA aims to transform from a space agency of 19 member countries into a space agency for the European Union.[7]

The Rising Sun

Japan Aerospace Exploration Agency

As the pace for activities in space was set by the Soviet Union and the US, the new players in the second wave of countries that entered the game thereafter did so to capitalise on the potential that was on offer. What was remarkable was that the second wave comprised three countries from Asia[8]; Japan's satellite launch in February 1970 was followed two months later by China, and then by India, in 1980.

Japan's space endeavours are an example of a nation initiating its space programme purely for peaceful purposes. It traces its origins to the pioneering work done by Professor Hideo Itokawa of Tokyo University. During World War II, Itokawa was a designer for the Imperial Japanese Army and an engineer with an aircraft company. It was Itokawa who campaigned for the Japanese government to participate in the International Geophysical Year (IGY) resulting in 13 launches of the Kappa-6 sounding rockets reaching an altitude of 60 km, during the IGY. In 1964, Japan established its Institute of Space and Aeronautical Science (ISAS) at the University of Tokyo, maintaining its close affiliation to the academia, operating largely in an academic environment. Four successive failures between 1966 and 1969 to launch Japan's first satellite led to the media ostracising Itokawa, who finally resigned. Establishment of the National Space Development Agency (NASDA) in October 1969 was undertaken to take the lead in developing space capabilities for peaceful purposes, and it was made responsible for developing satellites and launch vehicles. While ISAS retained its academic culture, NASDA evolved from more of an applied engineering environment. The same year, Japan and the US signed an agreement on space cooperation allowing for transfer of unclassified launch technology from the US to Japan, while, at the same time, prohibiting Japan from reexporting this technology to others.[9] This led to the first successful launch of the satellite, OSHUMI, in February 1970, and placed Japan as the fourth country in the world and the first Asian nation to do so. The launch site was the Kagoshima Space Centre and the ISAS-developed solid-fuelled Lamda-4S rocket was fired eastwards across the Pacific Ocean. The three-stage N 1 launcher developed by NASDA, a modified version of the McDonnell-Douglas Delta launcher, achieved success in 1975, with almost complete technical assistance and hardware from the US. It was the N 1 that launched Japan's first geostationary satellite, the Engineering Test Satellite (ETS) II, in February 1977, while the US Delta launcher placed the first of a series of Broadcast Satellites (BS) in GEO in 1978. The Japanese indigenous content in these two satellites was about 40 percent and 15 percent respectively, and 24 percent in the first communication satellite launched in 1978.[10]

Work on increasing the indigenous content in space capabilities commenced in the early 1980s as it would make Japan self-reliant in launching capability and be the harbinger of Japan's entry into the international commercial market. ETS IV, launched in 1981, was the country's first indigenously developed communications satellite and in the context of indigenous launchers, the H 2, a launch vehicle capable of placing a two-ton payload in GEO, took flight in 1994. The late 1990s witnessed a string of failures in H 2 launchers, forcing Japan to abandon further launches of the H 2 and concentrate on a reliable lower-cost production and operational launcher, the H 2A (H 2 Augmented), with the aim of beating international competition from Europe's Ariane, the Chinese Long March and then later from the Russian Proton launchers.[11] The first of the H 2A was launched in August 2001.

It was in October 2003 that Japan's space programme's consolidation received a major fillip. ISAS, NASDA and the National Aerospace Laboratory (NAL), working under different ministries of the government, were merged into one common agency, the Japan Aerospace Exploration Agency (JAXA). Not only was the aim to develop a unified culture and organisational structure for the country's space programme, but the move was also expected to save 10 billion yen a year.[12]

The H 2A launcher has proved to be Japan's mainstay, the space-workhorse launcher in the past two decades, whose simplified design and improved efficiency of the manufacturing and launch processes achieved one of the highest performance-to-cost ratios of launch systems in the world, reducing the cost of launches by a half or more.[13] This launcher was responsible for launching many of the satellites for information gathering, multipurpose transport systems, Earth observation, data relay and communications, astronomical research, asteroid exploration and also a lunar orbit explorer. An upgraded version of the H 2A, the H 2B, carried the country's launch prowess further when it launched the 16.5 ton H 2 Transport Vehicle (HTV) to the ISS in 2009. Japan has to its credit the distinction of placing the world's first solar powered sailcraft in space. IKAROS (Interplanetary Kite-craft Accelerated by Radiation Of the Sun) in operation since October 2011, relies on solar cells on a square shaped sail 20-m in diagonal length to propel the craft by generating electricity from sunlight on the solar sail.[14] Developments of the past two decades, being purely indigenous, have allowed Japan to leverage itself out of the limitations of the agreement signed with the US in 1969 and has, of late, launched its satellites with payloads developed by some of its international partners.

Japan's Handshake

Being a signatory to the ISS project since 1985 as also the second largest financial contributor, Japan has contributed to this mega project with its experimental space-lab module, KIBO, in March 2008 and has had, and continues to have, its astronauts aboard the space station. In addition to a select few countries that logistically sustain the ISS, Japan joined this core group in 2009 by launching the H 2 Transport Vehicle (HTV) named Kounotori, an unmanned and expendable transport spacecraft capable of ferrying loads up to six tons to the ISS and back, rocketed by the seasoned H 2B.

In the arena of international cooperation and collaboration, Japan has clearly broadened its horizons since the early days of its reliance on American support. With a cooperative exchange of information agreement signed between Japan and ESA as early as in 1972, the years since then have witnessed an increase in Japanese involvement with European partners: mutual support with France in the fields of launching Earth observation satellites and space radiation observation experiments; with the Swedish Space Corporation in establishing a mobile tracking station in Sweden and in the study of space sciences; 15 with Italy in the field of satellite applications for disaster management and a joint project for propulsion systems, including a hybrid space engine; and 16 with the Netherlands in the field of X-ray astronomy. An agreement with Russia enabled Japanese astronauts to board the Russian space station Mir. An agreement in 1989 between the space agencies of Canada and Japan led to mutual cooperation in the areas of Earth observation and microgravity technology. It was in 1993 that Japan spearheaded the establishment of the Asia-Pacific Regional Space Agency Forum (APRSAF), a platform where space agencies, governmental bodies and international organisations such as the United Nations, as well as companies, universities and research institutes from over 30 regional participants take part, working towards international cooperation in the Asia-Pacific region, focussed mainly in four distinct spheres of space activities: Earth observation, communications satellite applications, Space Education and Awareness (SEA) and Space Environment Utilisation (SEU).[17] Cooperative research has been undertaken by Japan with some of the space players like Australia, China, India, Indonesia and Korea.

The Road to Militarisation

The beginnings of the Japanese space programme lay essentially within the ambit of exploitation of space for peaceful purposes. Japan's post-war Constitution was drawn up under the Allied occupation that followed World War II, and its Article 9 is largely characterised by the renunciation of the right to wage war, as also the threat or use of force as a means of settling international disputes and a ban on maintaining a military force. The immediate years following this Constitution witnessed international events in the subcontinent that were to change this mandate. The post-war rivalry between the Soviet Union and the US had by now commenced in right earnest. Chiang Kai-shek, America's war-time ally in China, was defeated by the Communists and fled to Taiwan in 1949, and in 1950, the Korean War pitted the US against Communist forces in the Far East. As these events unfolded, the US came to see Japan as an important ally for the defence of US interests in the Far East and, ironically, in 1952, impressed upon them the need for rearmament. Japan created a National Police Reserve, a paramilitary force of 75,000 to defend the Japanese islands, which, at the behest of the Japanese Parliament, the Diet, evolved into the Self-Defence Forces (SDF) in 1954, a purely defensive military force numbering 270,000, a relatively small force compared to those of its neighbours.[18] While its charter remained that of maintaining the three wings of the military purely for the country's defence, the space programme differed slightly during its development, in that, the *programme had no military objectives* and was aimed solely towards peaceful purposes. By 1969, Japan had commenced buying satellite imageries for intelligence gathering from the US. These were not only expensive but also of inadequate resolution, further compounded by the fact that Japan did not possess the requisite capability in image interpretation. Indigenous satellites like the JERS satellite with optical and synthetic aperture radar, designed purely for commercial applications provided 18 m resolution, totally unsuitable for military reconnaissance and intelligence purposes.

This changed when North Korea launched a ballistic missile in August 1998 on a trajectory that overflew Japan, prompting Japan not only to participate in the US anti-ballistic missile programme, but give in to the unanimous support in the Diet for development of a satellite for intelligence gathering. The US in September 1999 offered participation of US firms in the programme as suppliers of sub-systems and components, but Japan chose not to buy a complete satellite from the US.[19] The Cabinet Satellite Intelligence Centre (CSIC) was established in 2003 to run the Japanese satellite reconnaissance programme with technical management by JAXA. The programme is centred around Information Gathering Satellites (IGS) whose mission is to provide early warning of hostile launches in the neighbourhood. Commencing with the launch of the first pair of IGS in March 2003, Japan has around half a dozen IGS in orbit providing resolution of about one metre, all launched aboard the H 2A rocket. A very large component of Japan's space-based military reconnaissance capability owes its success to the technical advancements achieved during the capability-development towards peaceful purposes. In addition, Japan has also placed in orbit a Quasi Zenith Satellite System (QZSS), a system that supplements the US' GPS. The higher latitude and mountainous topography of the country impinge upon reliable availability of the GPS signal. GPS signals are only available about 90 percent of the time in Japan, but satellite navigation will be possible 99.8 percent of the time with the QZSS underway.[20] As Japan's geographic latitudinal position does not afford placing a satellite in geostationary orbit above the country, the QZSS at an inclined orbital plane of about 40 degrees,[21] banks upon a number of such satellites at near-zenith over Japan, providing a sub-metre positional accuracy. The system not only enjoys tremendous commercial usage complementary to the GPS, but is dynamically scalable to provide military communications by equipping them with communications and broadcast sub-systems.[22] It is a matter of conjecture that need-based upgradation could provide QZSS the capability of exploitation for targeting purposes, that too at a time when the Diet in 2008, for the first time, allowed its military the use of space.[23] The Ministry of Defence (MoD) has since then been allocating funds for opening up a Space and Maritime Security Policy Office, with a mandate to study ways the MoD can utilise space and plan for future satellite systems.[24]

The Years Ahead

While for obvious reasons not much is known about its space-based military programme, what is a certainty is the uncertainty of the geopolitical environment in Japan's neighbourhood. A bellicose North Korea with a substantial space and missile capability, on the one hand, and a rising superpower, the belligerent China, on the other, will require Japan to relook at its options for national security. It could well be that its reconnaissance satellite programme and an increasing role in the US led theatre missile defence is a part of the growing willingness by Japan to play a bigger military role in the region.[25]

In the arena of programmes for peaceful purposes, Japan, armed with an annual space budget of US$ 3.8 billion,[26] has plans for building upon the technological capability proven so far in almost all the spheres of space activity. Japan plans to enter the commercial launch sector with its next generation Epsilon rocket launcher with reduced launch costs. The 91 ton three-stage solid-fuelled rocket aims to launch satellites up to 1,200 kg into LEO with an overall improvement in cost performance by 25 per cent.[27] The first of these rockets was launched in September 2013 which orbited a remote observation telescope at an altitude of 1,000 km. Towards Earth observation and meteorology, under development are satellites for measurement of global precipitation and global change, and Earth cloud, aerosol and radiation exploration. In the field of technology development, Japan continues to develop and refine small satellites under its Small Demonstration Satellite (SDS) project. With a 50 kg SDS 4 next slated for launch, the SDS programme launches small satellites with an aim to verify new technologies for components and devices in space prior to providing highly established technology to practical satellites and scientific probes.[28]

The country has plans for beyond-Earth missions too. Taking off from the successful SELENE, or Kaguya, mission which was launched aboard the H 2A in 2007 and landed a platform on the Moon in June 2012, the next logical step is going beyond the Moon. For inter-planetary exploration, a joint JAXA-ESA spacecraft, the BepiColombo, aboard the Ariane-5 is planned for the study of Mercury's surface and its magnetic field.[29] A spacecraft, Hayabusa 2, is planned to be launched in 2014, taking another four years to reach the targeted asteroid and get back to Earth by 2020.[30] This is a follow up to the ISAS launched Hayabusa in 2003 that landed on an asteroid named Itokawa and returned to the Earth in June 2010 with 1,500 tiny pieces of the asteroid.

Japan, as it stands today as the most technologically accomplished space power in Asia,[31] sees the years ahead with optimism, aiming to consolidate the position it has acquired as a responsible and significant space player. While Japan's launch services remain more expensive as compared to those of India, China and Russia, reliability and political factors will also step in to play a role in such decisions and Japan is likely to a have a small but slowly expanding commercial market, driven by the fact that continued growth in space capability remains critical to its image, technological reputation and security.[32]

The Dragon in Space

China's Rise in Space

October 15, 2003, went down in the annals of Chinese history as a major milestone achieved when China put its own *taikonaut*, the Chinese term for an astronaut, into orbit. Yang Liwei, a 38-year-old pilot of the nation's air force orbited the Earth 14 times aboard the spacecraft Shenzhou 5 launched by the Long March 2 rocket, and made a safe reentry and landing 21 hours later. The milestone that was achieved was China emerging as the first nation other than the two superpowers, to do so. Among countries other than the US and the Soviet Union that planned to launch their own first-satellite-in-space, China was clearly bested by France and then by Japan but it was the successful culmination of an eleven-year-long project to launch its own astronaut that established China as an emerging space player at the global level and a contender for the status of Asia's leading spacefaring nation. A brief look at China's build-up in space capabilities actually indicates that China had the potential to reach there earlier but for certain domestic compulsions that later unfolded.

The early years of the People's Republic of China's (PRC's) space efforts received a contribution in some way or the other from both the Soviet Union and, ironically, the USA. The US' unwitting participation came about in the form of a Chinese engineer, Tsien Hsue Shen (or Qian Xuesen), born in Zhejiang, China, in 1911, who migrated to the US for higher studies in 1935. Tsien earned his Ph D in 1939 from the California Institute of Technology (CalTech) under the mentorship of the famous Theodore von Karman. His work on amateur rocketry drew interest from the US military and by 1942, Tsien was working on small solid rockets. He became one of the co-founders of the Jet Propulsion Laboratory (JPL), the laboratory that CalTech operates for NASA, and the first head of research analyses at JPL in 1944. In May 1945, with an honorary rank of Colonel of the US Air Force, Tsien landed in Germany and acquainted himself with Wernher von Braun[33] and his work. As a professor of jet propulsion in CalTech, he predicted manned travel to the Moon within the next 30 years and some of his ideas inspired the US Air Force (USAF) to develop one of the ancestors of the Space Shuttle.[34] And after all the expertise gathered and contribution made in the field of rocket design for the US, Tsien in 1951, accused of being a Communist was placed under house arrest and deported, along with other Chinese scientists in 1955 during the peak of the anti-Communist witchhunt that plagued the US in that era.

The US' loss was to be China's gain as Tsien reached his motherland to a warm welcome. Armed with up-to-date theory of rocketry in his head, the theory that would translate in the next few decades into Tsien achieving the status of Germany's von Braun, the Soviets' Korolov and America's Goddard, the first unwitting step towards space had been taken for China. China in the early 1950s was in a state of flux, moving from a state of social unrest, the war with Japan and the Communist revolution to a state of socialist reconstruction and modernisation. Tsien's presentation in 1956 on the establishment of China's defence aeronautics industry was well received by the Central Committee of the Communist Party and paved the way for the establishment of the Fifth Research Academy on October 8, 1956, to spearhead China's space programme, with Tsien as the first Director of its Rocket Research Institute.[35] That the existence of the Fifth Academy was classified and placed under the Ministry of National Defence is indicative of the fact that China's first institutionalised step towards space was clearly aimed at military pursuits. The very name 'Fifth Academy', too generic to give away its actual purpose, also points to an attempt towards secrecy, akin to China's Ninth Academy which was responsible for nuclear weapons development.

While Tsien provided the theoretical stewardship to the nation's space ambitions, the nation by itself did not possess the basic requisites of rocket hardware. This was the era of the Cold War in which the two superpowers had long range bombers as well as ongoing projects for ballistic missiles for the delivery of nuclear weapons. China had neither, and viewed in the backdrop of the American threat of a nuclear attack at the end of the Korean War,[36] it took to capability development in the nuclear programme and in ballistic missiles as the means for their delivery. To avoid commencing work on rockets from scratch, China turned to external support and it was at this point that the Soviets' contribution came in. The period from 1956 to 1958 saw the arrival of the Soviet R-1 and R-2 rockets along with technicians and documentation under a cooperative technology transfer programme. The Sputnik 1's launch in October 1957 prompted China to plan for a satellite launch in addition to its ongoing rocket development with a view to spur the development in other fields of science and technology, and mobilise military rocket development.[37] However, a collapsing economy and nationwide starvation in the late 1950s forced China to shelve the satellite project temporarily and focus on sounding rockets and military imperatives of rocket technology. The breakdown of Chinese-Soviet relations in August 1960 left China feeling isolated and friendless, and further strengthened its belief in prioritising missile development over satellite projects.[38]

China's split from the Soviet Union occurred at a time when its development of the Dong Feng (DF) 1, the Russian R-2, was in an advanced stage and successfully flew for the first time in November 1960. Buoyed by the success of the sounding rockets and ongoing progress on the DF series of military missiles, Tsien sought and received the go-ahead of the Central Committee in 1965 for a Chinese satellite. Thus, after a hiatus of six years, the first Chinese satellite, the Dong Fang Hong, was launched successfully on April 24, 1970, enabling China to becomer the fourth country to do so. The launch vehicle employed was the Chang Zheng 1 (CZ 1), or the Long March 1 rocket, a civil version of the DF 4 MRBM. The subsequent CZ 2, based on the ICBM DF 5,[39] and the series of Long March rockets, would thereafter continue as China's space workhorse for forays into space in the decades to come, including offering commercial launch facilities to partner states.

White Papers on Space

China publishes White Papers on various fields of activity at the national level, including defence, with an aim to offer its own people and the international community at large an insight into government level policy issues. Being a Communist state, the White Papers[40] accord a certain amount of transparency in the functioning of the state and aim towards confidence building in the world community of China's intent. The PRC has published three such papers on space in the years 2000, 2006 and 2011. Each of the three papers covers activities undertaken in space in the previous intervening years and lays out a broad plan for the next five years. The early space related activities were by and large kept under wraps and not much information was forthcoming from China into the public domain. Whatever little was known to the outside world came about through surveillance resources of the USA. As early as in 1958, U-2 reconnaissance aircraft and then Discoverer/Corona photo-reconnaissance satellites were engaged in keeping watch over China's nuclear and military programmes.[41] The first documented account of China's space programme from China, was the first White Paper on space published in 2000, which bridged the gap in information and confirmed to the world the status achieved by China. The White Papers enunciate the principles that govern China's space programme, as:[42]

- Adhering to the principle of long-term, stable and sustainable development and making the development of space activities cater to, and serve, the state's comprehensive development strategy.
- Upholding the principle of independence, self-reliance and self-renovation and actively promoting international exchanges and cooperation.
- Selecting a limited number of targets and making breakthroughs in key areas according to the national situation and strength.
- Enhancing the social and economic returns of space activities and paying attention to the motivation of technological progress.
- Sticking to integrated planning, combination of long-term and short-term development, combination of spacecraft and ground equipment, and coordinated development.

The aspect of the PRC's military objectives and endeavours in space is conspicuous by its absence. The three White Papers advocate use of space for peaceful purposes, the role that the United Nations has to play and China's participation in the charter of the United Nations Committee on the Peaceful Use of Outer Space (UNCOPOUS).

Civil Space Programme

Early steps towards space by China may have had their roots in the cooperative agreement with the Soviet Union, but 1960 onwards, the breakdown of ties between the two nations forced the PRC to go its own way. It did so, albeit slowly and doggedly, encompassing all space activities in the build-up and expansion of its national power. Steered by visionary leadership and single-minded ambition, the space programme developed, encompassing all facets of the space spectrum. Extensive innovation-led research and development has contributed to a large indigenous manufacturing base that spans the entire spectrum from building satellites to producing launch vehicles. It is, however, in the field of satellites, including dual-use ones, that China has sent a global signal of 'having arrived' as a rising power with the potential of being a global game-changer.

The three decades since the launch of China's first satellite in April 1970, saw 47 satellites of various types being successfully placed in orbit. These essentially were of four types: the Fanhui Shei Weixing (FSW) series of recoverable experiment satellites, the Dong Fang Hong (DFH) series of telecommunications satellites, Fengyun (FY) meteorological satellites and the Shijian (SJ) series of scientific research and technological experiment satellites. In addition, by the mid-1980s, China had commenced using the communications satellites IntelSat and InmarSat, of other countries and developed related technologies. The deployment of communications satellites by the mid-1990s had reached a stage where it could provide coverage to 83 percent of the Chinese population.[43] With the establishment of the China National Space Administration (CNSA) in 1993, a government institution to develop and fulfil China's due international obligations,[44] one of the first joint collaborative Earth observation satellite projects, the CBERS (China-Brazil Earth Resources Satellite) fructified with the launch of the CBERS 1 in 1999, aboard the Long March 4B rocket and then with CBERS 2 in 2003.[45] This project was a take-off for the fifth category of satellites, China's indigenous Earth observation satellites, the Zi Yuan (ZY). Subsequent refinements in the ZY series of satellites now provide three dimensional, multi-spectral observation of the globe.

The turn of the century opened up two symbolic chapters to China's growth-in-space history. Not necessarily appearing in chronological order, was the technology demonstrated in the field of microsatellites (microsats) of up to 100 kg and nanosatellites (nanosats) of up to 10 kg mass.[46] Seeing the potential of small sized satellites in civil as well as military applications, China commenced work on microsats in the late 1990s. In participation with the UK-based Surrey Satellite Technologies Limited (SSTL) commencing 1998, China's first microsat, the 50 kg Tsinghua 1, was launched in June 2000, providing a 39 m resolution,[47] to conduct research in low Earth orbit, as well as being a demonstrator for an eventual constellation of five Chinese microsats intended to provide global, high resolution imagery.

Its successor, the Tsinghua 2, or Beijing 1, launched in October 2005, provides a 600 km area coverage with a 4 m resolution,[48] which would have a significant military implication. By this time, China had gone ahead with further iterations and launched a 35 kg microsat in April 2004, firmly establishing its toehold in this niche segment of space technology. The obvious implications of developing and fielding such a technology are two-fold. Firstly, it provides the country ample scope for employing such satellites for developing and evaluating new applications in the space environment, prior to placing in orbit a fully developed operational space asset. The second fallout leans towards the military wherein it permits the flexibility of 'launch-on-demand' – placing in orbit a payload that meets an on-the-spot operational requirement. The ready-to-launch satellite, of course, would need an equally ready-to-fire launch vehicle. China has addressed the latter issue by pursuing development of small-lift solid-fuel road mobile rockets. The four-stage Kaituozhe 1 (KT-1) launch vehicle designed to place a 50 kg satellite in LEO[49] has made three test flights since 2002. The same, though not in operational use, accords China the flexibility of launch-on-demand, giving the country a 'reactive' launch capability, that is, a capability to place satellites and space launch vehicles in storage, to launch within 24 hours[50] of a decision to do so.

The second major milestone in this millennium came about in the field of GPS. With no forewarning, the first of a series of Beidou navigation satellites took to the skies in October 2000. Operational since December 2011, the Beidou regional positioning satellite system, with satellites with a mix of MEO and GEO orbits, provides 10 m accuracy for China and its immediate neighbourhood.[51] This navigation system then will graduate onto the Compass Navigation Satellite System which aims for global coverage with 35 satellites by the year 2020.[52]

The early decades of the Chinese space programme did not lay any stress on manned space flight activity. Human space flight, conventionally, does not accrue any military or economic dividend and this was the view of the Chinese leadership too. Having proven its status as a major space player over the years, the thought of nationalist pride and international prestige with 'techno-nationalistic' overtones, however, swung the focus towards attempts at a manned programme. Thirty-four years after the estrangement between the Soviet Union and China in 1960, the ties were reestablished[53] and in 1994, Russia sold some of its advanced aviation and space technologies to China. One such technology led to the Shenzou manned space flight programme in China. The manned space programme, codenamed Project 21, was started in 1992 and has seen US$ 6 billion spent on it, about half of which was spent on developing advanced technologies for rendezvous and docking.[54] It was the fifth in the series of Shenzou[55] spacecraft that carried China's first *taikonaut* into space in October 2003 while Shenzou 7 facilitated the country's first spacewalk in September 2008. The step-by-step approach towards space led China to launch its first space laboratory module, the Tiangong 1, aboard the Long March 2F/G rocket in September 2011. This is an experimental orbiting platform aimed at honing docking techniques required to support a full-fledged space station in the years to come. As the next natural step, an unmanned Shenzou 8 spacecraft docked[56] with the Tiangong 1 twice in November 2011. The docking was carried out in an automated manner, as a precursor to a manned docking by the Shenzhou 9 that docked with the Tiangong 1 on a fortnight long mission in June 2012. This mission was unique in that its crew included China's first woman *taikonaut,* Liu Yang, a 33-year old air force pilot in charge of medical experiments during the mission. The Shenzhou 10, in June 2013 carried a three-man crew. With these three successful missions, docking techniques have been honed and the Tiangong 1 is ready to be deorbited.

Not quite content with activities in the space around the Earth, China emboldened enough, pursued exploration beyond the Earth. The lunar probe Chang'e 1 in October 2007 orbited the Moon, making China the fifth country in the world to achieve this, and was followed by the Chang'e 2 in October 2010, providing China with a complete high-resolution map of the Moon. The next step in showcasing Chinese deep-space capability was logically a probe to the planet Mars. The YInghuo 1, a Mars probe aimed to orbit its Moon, the Phobos, was launched in November 2011, from Kazakhastan, along with a Russian return spacecraft aboard the Russian Zenith 3F rocket, but, unfortunately, was unsuccessful and made a destructive reentry two months later.

Launch Vehicles

The success achieved by China in the field of its satellites has been complemented with an equally successful record of its launch vehicles. Humble beginnings, with assistance from the Soviet Union and then the breakdown in the relations between the two, necessitated China to have a detailed look and review how it had to go ahead if it wanted to have a say in the world polity. The route to China's first Long March 1 rocket that launched its first satellite, originated from the German von Braun's designs and reached the Chinese mainland through the Soviet R-2 rocket. From 1960 onwards, China consolidated what it had gathered and commenced its quest for its slice of the space-pie. And it did so rather remarkably, despite the fact that its odyssey in space was a byproduct of its pursuit of matters military. The series of long range Dong Feng (DF) missiles, developed for military payloads, including nuclear, paved the way for a series of correspondingly-capable civil-use Chang Zheng (CZ), or the Long March rockets, as they are commonly known.

It was the Long March 1 that launched China's first satellite in 1970 and Long March 2, derived from the ICBM DF-5 that launched the FSW recoverable satellites commencing 1974. As it stands today, China's space launch capability rests on around a dozen different configurations of the Long March 2, 3 and 4 rockets, capable of launching different payloads, of various tonnages to different orbital altitudes, including to the Moon and beyond. The Long March 2C and 2D are solid-fuelled two or three-stage medium lift launchers used for placing 3-4 ton payloads in LEO. With four strap-on liquid-fuelled boosters in addition, the Long March 2E places up to 4.8 ton payload in geosynchronous orbits while the 2F is employed for manned space flights in LEO. The Long March 3 places DFH satellites in GEO and Long March 4 has been exclusively used for placing over 2 ton payloads in Sun-synchronous and high-inclination polar orbits.[57] The technological advancements in the field of successful launches can be discerned from the fact that in the year 2011, China surpassed the US with 18 successful launches out of a total 19 as against 17 from 18 by the US.[58] Statistics for 2012 indicate 19 successful launches for China as against 12 successful out of 13 for the US, making the Long March family the most oft-flown space launcher in the world.[59]

Launch facilities exist at three sites in mainland China, one each exclusively for LEO, polar and GEO missions. All three, though modern, are landlocked and do not provide the benefit of launches from lower latitudes. China has chosen Wenchang, in the island province of Hainan near the southern end of the mainland, as its next launch site. At a latitude of 19 degrees north of the Equator and with the South China Sea on its east, the site offers tremendous advantage of launching eastwards, close to the Equator, allowing for a substantial increase in payload, so necessary for the future manned missions, space station and deep space exploration programmes that China aims to undertake in the years to come. An island based site assumes added significance since the next generation of Long March 5 launchers are aimed to be 5 m diameter core heavy-lift rockets, which perforce would necessitate transportation by sea from the construction site at Binhai, a port city near Beijing. Wenchang, a former sub-orbital test centre is currently undergoing major upgrades to make the fourth space-port a modern high technology facility.

China's Military Initiatives

> *China is threatened by US policies in space, a reality that is compelling China to have its own space capabilities...A deterrent in space will decrease the possibility of the US attacking Chinese space assets...Chinese deterrent policy in space will vigorously maintain the 'active defence' strategy.*[60]
>
> — Bao Shixiu
>
> Chinese Military Scholar
>
> People's Republic Army Academy of Military Sciences

Were the above words a justification of China's first-ever overt aggressive display of military might in space? On January 11, 2007, China launched a ground-based MRBM SC 19 and destroyed its own defunct weather satellite, the Fengyun 1C, in a polar orbit at an altitude of 865 km. The Anti-Satellite (ASAT) test did not violate any existing arms control treaty but it did break a voluntary moratorium since the 1980s on such destruction of a satellite[61] and created over 2,700 pieces of trackable pieces of debris (larger than 10 cm) and likely thousands of smaller pieces.[62] This was the first known

ASAT intercept after the US' air-launched ASAT kill of its P78-1 satellite in 1985, leading to speculation worldwide on the timing of such a display. China, along with Russia, had repeatedly advocated an agreement to ban weapons in space by proposing the reestablishment of an ad hoc committee on the Prevention of an Arms Race in Outer Space (PAROS) at the United Nations' Conference of Disarmament (CD). The proposals had not found favour with the US that wanted to retain its flexibility for developing its anti-missile defences[63] and, for the first time, had voted against the treaty to ban weapons in space in 2005, as against abstaining to vote. The hegemonic practices by the US further compounded the issue when the US National Space Policy of 2006 laid out aggressive plans for dominance in space that included protecting its space capabilities and denying adversaries the use of space capabilities inimical to US interests. In Michael Krepon's words, the ASAT test was "...a predictable and unfortunate response to US space policies".[64] Not just an act against the US' hegemonic attitude, the ASAT test of January 2007 duly confirmed to the world at large what hithertofore it had suspected—within the gamut of exploiting space as a medium for projecting its national power, China had a credible capability in space for its national military pursuits. What came to the fore was the fact China was not only equipped with 'force enhancement' or support capability in space, but now possessed demonstrable 'space control' or offensive capability.

The three White Papers published on space by China, for obvious reasons, do not pointedly mention any military content in its space programme. However, the White Paper on "China's National Defence" published in 2006 establishes the linkage between space and national defence. Section VIII on 'Science, Technology and Leadership for National Defence' has the words "...priority will be given to upgrading technologies and products in the nuclear, *space*, aviation, shipbuilding, weaponry, electronics and other *defence-related industries*.." that more than hint at the role space plays in China's security strategy.

While China may have stated the role envisaged for space in its national defence in 2006 and demonstrated that capability a year later by its ASAT test, China's space programme has always been "inherently military in nature... indeed, it is a military-civil joint venture in which the military develops and operates its satellites and runs its infrastructure, including launch sites and satellite operations centre."[65] It was the military that kept the missile and space launcher projects insulated and on course, while major upheavals tore the Chinese society in the 1950s (the anti-right campaign and the Great Leap Forward) and then again in the 1960s (the devastating famine and the Cultural Revolution).[66] While the 1960s saw the DF missile development paving the way for Long March civil-use launchers, the 1970s saw Chinese satellites of various roles take to the skies and it was the 1980s that witnessed China's fielding of dual-use spacecraft. Moltz, a faculty appointment in the US' Department of National Security Affairs, argues that China's military space operations received a significant impetus on witnessing the USA's expert use of space in the 1993 and 2003 Gulf Wars, given the omnipotent role it played in support of large-scale modern military operations.[67] Thereafter, China took to developing and launching satellites dedicated for the military while continuing with the civil-use satellites. Whatever little has come out of China on the issue of techno-military capabilities of its satellites, is obfuscated with civil-use purpose and nomenclature that could be applied to various meanings and interpretations. Dual-use technology further compounds classifying and cataloguing the space assets, for example, the satellite ChinaSat 22, launched in 2006 and catalogued as a communications satellite, is believed to be a military satellite named Feng Huo 2, a part of the data-relay network system that enables theatre commanders to communicate with other forces.

Among the various constellations of satellites that China operates, three have very definitive military roles: the FSH, Yaogan and Tianhui,[68] all of which are operated by the People's Liberation Army (PLA). The FSW series of recoverable satellites commenced operations in 1975. Their military designation was JianBing which launched 24 recoverable reconnaissance satellites and successfully recovered 22. The FSW satellites were designed to capture low- to medium-resolution (10-15 m) images in visible light on photographic film, and then return the film to Earth inside its reentry capsule. The FSW series of recoverable reconnaissance satellite project was terminated in 2005 as China progressed to transmitting imagery to ground control in near real-time. The Yaogan (YG) series operates over a dozen 'remote sensing' satellites whose military designation is also JianBing (Jianbing 5 onwards).

Commencing operations in 2006, with orbits at 600-1,100 km altitude band in LEO, these satellites are believed to be in use for Intelligence, Surveillance and Reconnaissance (ISR) roles. Despite carrying the same designation, the YG satellites actually include several different variants in two main categories: Synthetic Aperture Radar (SAR) all-weather surveillance satellites and electro-optical digital imagery satellites. The L band SAR aboard the reconnaissance satellite, the YG 1, provides a 5 m resolution over a 40 km swath, while the YG 2 is an electro-optical digital imaging satellite with a 1.5 m resolution. Each successive YG satellite achieved enhanced capability and it was the YG 9, launched in March 2010, that was built around a constellation of three satellites. The YG 9A, 9B and 9C satellites comprising an electro-optical surveillance satellite, a SAR satellite, and the third, possibly, an Electronic/Signal Intelligence (ELINT/SIGINT) satellite, are designed for locating and tracking foreign warships, in particular aircraft carrier battle groups, by collecting the ships' optical and radio electronic signatures. Combined with other systems such as maritime surveillance satellites and data relay satellites, the YG 9 constellation provides an over-the-horizon reconnaissance and target designation capability for China's maritime strike forces, such as the Anti-Ship Ballistic Missile (ASBM) system. The SAR satellites have special military significance as SAR has the capability of tracking moving objects and seeing through clouds, smoke and fog during day and night operations. The Tianhui series of surveillance satellites operate in LEO and their CCD cameras provide 5 m resolution.[69]

Besides the above, CBERS, the joint venture between China and Brazil which enabled China to hone its indigenous satellite development, led to the Earth observation ZiYuan series of satellites. The early ZiYuan satellites evolved into military-enabling ones like the ZiYuan 3, launched in January 2012, that provides a 2.5 m optical resolution and a 10 m resolution by its infrared multispectral scanner,[70] and has the capability of transmitting images to Earth in near real-time. This satellite meets international standards, ridding the country of its reliance on imports of satellite imagery.[71]

Besides military satellites operating in the domain of reconnaissance, China lays equal stress on the realm of communications satellites for the military. This, in part, is necessitated by the country's stated objective of "winning local wars under an informationalized environment". Armed conflicts around the globe in the last few decades bring to the fore the need for Network Centric Warfare (NCW). Intrinsic to the concept of NCW is the role that communications and data-handling play. Enhanced bandwidth with in-built redundancy is an issue that China has commenced addressing, in both the civil and military spheres. With respect to the latter, the military version of the ChinaSat series of communications satellites was named Feng Huo (FH), the first of which was launched in 2000. The FH tactical communications satellites in GEO, are designed to provide Communications-On-The-Move (COTM) on the Ku- and Ka-bands in VHF, which enable military users to send and receive communications and maintain situational awareness whilst in motion on land, in the air and at sea. Conventional satellite communications require the user to be stationary so that the antenna can be pointed towards the satellite, which can be a huge disadvantage in the battlefield. COTM eliminates this need to pause by having the antenna automatically tracking the satellite in motion.[72] This communications architecture is in place for the Qu Dian C4I system, China's first integrated Command, Control, Communications, Computer and Intelligence (C4I) system. The Qu Dian system gives the Chinese military new capabilities for coordinating and supporting its terrestrial forces, quite akin to the American Joint Tactical Information Distribution System (JTIDS). When fully deployed, the Qu Dian system will allow theatre commanders to communicate, and share data with all the forces under joint command. The TianLian series of tracking and data relay satellites, the first of which was launched in April 2008, gives China the benefit of extended coverage and control over its satellites in orbit, which has tremendous potential for exploitation and capability enhancement of military satellites.

Over and above the 'space support' activities undertaken by China in support of its military, it has also embarked upon the aggressive or offensive content of space-force applications, that is, the '(space) attack' or counter-space operations. While ISR satellites constitute the eyes and ears in space, and communications satellites facilitate communications, *killer satellites* constitute the other end of the space continuum. For an enemy's satellites to be attacked as part of 'counter-space' operations, locating them would need to be the first step. As a prerequisite to 'space situational awareness' so essential to any military campaign, China has an extensive network of tracking stations, including overseas, both land-based as well as ship-based. Its space-control network enables China to track its fleet of orbiting satellites with an accuracy of one metre.[73] This has enabled China to maintain an increasingly comprehensive catalogue of the relevant space objects.[74] The means of achieving this, besides the open-source material, have also been covert intelligence gathering[75] and a myriad range of optical telescopes, radars of air defence systems, phased array radars and various ground- and space-based signal intelligence systems. Having acquired the orbital parameters of the space objects and their spatial relationship with other orbiting objects, if the situation so arises, China has the wherewithal to actively degrade or engage or destroy the adversary's space platforms.

The ASAT test was a demonstrable technological achievement of a direct attack weapon, or a hard kill or kinetic weapon. The ASAT fired in January 2007 was one that can effectively destroy a space platform in LEO. Against satellites in GEO, several launchers and ballistic missiles like the new DF-31 ICBM are capable of carrying an ASAT to geosynchronous orbit, where the relative stationary location of satellites makes them vulnerable.[76] China also has an ongoing programme for a co-orbital ASAT weapon, which would be positioned into a temporary parking orbit from which it would manoeuvre to attack its specific target.[77] This capability was proudly broadcast by Chinese television when it gave the coverage of the spacewalk by the crew of the Shenzhou 7 in 2008. What caused consternation was the simultaneous release of a 40 kg satellite, the BanXing-1, which took pictures of the Shenzhou 7 and in the process, came within 25 km of the ISS.[78] Within the tactics of co-orbital weapons, microsats and nanosats are the most problematic[79] as ASATs, as they can quickly launched from mobile launchers or launched as secondary payloads from larger launchers carrying civil-use payloads, and are difficult to detect and track. Directed Energy Weapons (DEW) like ground-based lasers are also another means of hard-kill of orbiting satellites and it can safely be surmised that China would, or may have concentrated on this facet too. In addition, destroying the adversary's ground control infrastructure would also lie in the domain of 'hard-kill' options.

Counter-space operations need not necessarily bank upon the hard-kill option alone. Actions within the scope of soft-kill operations include a nuclear detonation in LEO, the resultant Electro-Magnetic Pulse (EMP) rendering inoperable the electronics aboard satellites in line-of-sight range. Another option lies in electronic jamming or 'blinding' the satellites by ground- or space-based jammers, with one variant of this technique being jamming of the uplink or the downlink between the satellite and its ground control itself. While only the ASAT test has been demonstrated and there exist unconfirmed reports in certain quarters of China's 'dazzling' or electronic illumination of American satellites overflying China by ground-based laser in August 2006, China's capabilities in hard- and soft-kill arenas are largely speculative, shrouded as they are in a cloak of secrecy. This cloak is opened every once in a while to the outside world, like the display of the ZM-87 laser weapon in 1995, at defence exhibitions in Manila and Abu Dhabi.[80] There is, however, little doubt that China is techno-politically capable of carrying out such operations but is unlikely to do so unless it is under extreme circumstances, as an act of desperation or self-preservation.

Cooperative Ventures

As compared to the two superpowers in the 1950s and 1960s, China is a relatively newer player in the field of space. Space endeavours being heavily technology-dependent, China had much to catch up with. With a poor and underdeveloped domestic military-industrial base, whose indigenous weapons were at least a decade and a half behind the Western contemporaries,[81] China took to agreements initially with the Soviet Union to kick-start its first step into space. The break-up in the relations between the two forced China to take the indigenous route but it was soon made aware that it simply did not have the technology inherent in such a field. While there may have been reports of Chinese attempts at stealing this technology, the prowess acquired over the years is, to a large extent, also a result of collaborative ventures with other countries.

In the international arena, China became a member of the United Nations Committee on the Peaceful Uses of Outer Space (UN COPUS) in 1980 and followed it up in the same decade by acceding to treaties governing the use of outer space, on astronauts and on liability for damage caused by space objects. It was in 1989 that diplomatic relations between the USSR and China were restored and yet again it was in the field of space that China sought technological assistance. Joint partnerships and ventures with other countries commencing in the 1990s aimed to incorporate their advanced technologies within China's domestic technology base. Ventures with Germany, gave the Chinese advanced communications technology and enhanced lifespan capability in the DFH satellites; with Brazil came the CBERS project, enabling China to develop its indigenous Zi Yuan series of Earth observation satellites and the subsequent JianBing series of reconnaissance satellites; a tri-nation project with Germany and France that launched the SinoSat 1 satellite paved the way for China's capability-enhancement in advanced commercial communications satellites; receiving imagery from Canada's RadarSat helped China focus its attention towards developing its SAR capability; and the tie-up with the UK led to China's foray into the world of microsats and nanosats.

The cooperative approach with its partners worked equally well in the field of launch capability. The Long March rocket, China's ubiquitous space horse with an enviable track record in reliability, serves the international market and has been providing satellite launches commercially since 1986. As of June 2013, China has launched 39 foreign satellites into orbit, including of the US. China's share of the global commercial launch industry is a mere 3 per cent that aims at 15 per cent by the year 2020.[82] The fact that the launch services are provided at very competitive rates not only makes this venture commercially viable for its foreign customers, but greatly enhances China's ever expanding 'soft power' and also boosts its techno-industrial status worldwide.

The Years Ahead

China continues to view space as a medium for its strategy of developing and maintaining its comprehensive national power. China aims to be a global superpower, with space related activities playing a predominant role as evidenced by the heavy investment committed to its space programme. Abundant in technological and financial resources, with an annual space budget estimated to be US$ 2.2 billion,[83] and backed by the requisite skilled manpower, China today stands poised to stamp its dominance in just about every facet of space activity. If it has intentions of challenging the US' hegemony in space in the coming years, a lot more investment in the space industry would be essential as the US' annual civil space budget, though declining slightly in the last few years, still stood at a mammoth US$ 18 billion in 2012.

As would be obvious, China has no stated plans for the military component of its space projects. On matters military, it can safely be conjectured that the 'superpower tag' will necessarily entail extensive developmental and capability-enhancement work in the entire spectrum of 'space safeguard', 'space support' and 'space attack' activities.

The civil applications field is one where the PRC has enunciated plans for some bold and ambitious projects. The Long March 5 rocket slated for launch in 2014, with a planned capability to place 25 tonnes in LEO and 14 tonnes in a geosynchronous orbit,[84] would make it the second largest launcher in the world after Boeing's Delta IV Heavy rocket. The Long March 5 will be a new generation rocket that employs low-cost, non-toxic and non-polluting propellants. China aims to launch this jumbo rocket from the launch centre at Wenchang, Hainan, where work is currently ongoing to develop the fourth launch centre, expected to be operational, again by 2014.[85] Equally ambitious is the plan for satellite launches in the coming years. According to Zhang Jianheng, Deputy General Manager of the China Aerospace Science and Technology Corporation (CASC) China aims to launch 100 satellites in the period 2011-15.[86] In 2012, China, with 19 satellite launches, surpassed the US' 12 successful satellites launches and trailed Russia's 24.[87] The 100 satellites would also include a variety of payloads aboard micro- and nanosats as well as satellites that cover the gaps and expand the Beidou and Compass navigation system.

In terms of its manned space programme, China plans to build upon the capability exhibited in its space laboratory, the Tiangong 1, currently in orbit. Tiangong 2, planned for launch in 2015-16, aims to provide facilities for a three-man crew for 20 days while Tiangong 3, to be launched two years after Tiangong 2, would support a habitat for 40 days. The aim of both would be to examine regenerative life-support technology, as well as to verify technologies of orbital replenishment of propellant and air. The overall aim is take the lessons learnt from the space laboratory project to a full-fledged space station. To be assembled in the 2020-22 time period, this large orbital station will weigh about 60 tons and with a design lifetime of about 10 years, will support three astronauts for long-term habitation.[88]

Plans are afoot even for missions beyond the Earth's gravity. China has planned for a robotic lunar sample return mission in 2017, followed by a manned mission some time in the next decade. Besides achieving a major milestone in the annals of Chinese national prestige, missions to the Moon could well be a necessity for human civilisation at large in the decades to come. As the world population swells to 12 billion by the year 2050, global energy needs will grow eight-fold.[89] Scientists view Helium 3 as the fuel source of the 21st century as it is extremely potent, non-polluting and with no radioactive byproduct. Just 25 tons of Helium 3 could meet the US' energy needs for one year and while there is hardly any Helium 3 on the Earth, there is an estimated one million tons of it on the surface of the Moon.

China's visibility in the space arena came to be seen in the 1990s in its launch systems and then more and more so in the 2000s in its satellites, manned mission and then in the military domain when it conducted its ASAT test. These last five decades of the Chinese space programme have established China as a frontrunner in all aspects of successful space research, development, production and launch activities, exhibiting an end-to-end capability from launch centres to launch vehicles, to satellites providing services across the entire spectrum of applications, space situational awareness to spacecraft recovery operations, to manned space missions and deep space projects. Equally active in the field of international cooperation and agreements, China has further strengthened its technology base and opened up its launch and manufacturing industry to commercial ventures. In matters of military-space, which way China will go remains to be seen as it is likely to be governed, not in isolation by internal domestic compulsions but in all probability by the global geopolitic use of space by the major space players in the times to come.

Israel's Space Ambitions

The Geostrategic Compulsions

If there is a country in the world today whose military pursuits have been driven only by the intense geopolitical and geostrategic compulsions that challenge its national security, it is Israel. Surrounded on three sides by the Arab world, Israel had little choice but to invest heavily on its techno-military industry. The small size of the country, lack of geographic depth and the omnipresent threat from the region formed the cornerstones of Israel's vision for its national security, and 'survival of the state' remains the lynchpin in the country's security paradigm. Building its military might had technology and innovation as its key ingredients. In the words of Steven Hunt "...I've always found that some of the freshest thinking about security comes out of Israel. I suppose that seems obvious. After all, security is a part of the fabric of daily life in that little country".[90] Israel ranks as one of the leading global players in terms of scientific research, science publications and technology development.

Israel entered the elite club of 'spacefarers' on September 19, 1988, with the launch of an indigenously built 156 kg demonstration satellite, the Ofeq 1, aboard its own three-stage, solid propellant rocket, the Shavit, making the country the eighth in the world to do so. Israel had achieved this in a manner similar to that of its predecessors other than Japan and India, in that, military objectives were a precurser to exploits in space and military missiles provided space launchers as a byproduct. What is unique in the case of Israel's launch is the geographical handicap of the country's location. Countries in the east inimical to its interests preclude Israel from launching to the east. The Shavit was launched in a northwestern direction over the Mediterranean Sea, and continues to do so, with the resultant disadvantage of orbiting satellites of reduced weight as the direction of launch does not harness the advantage of launching in the direction of the Earth's rotation. The Shavit launcher is believed to have been developed from the Israeli Jericho 2 long range missile,[91] and by some reports "the Jericho 2 is a Shavit minus the upper stage, which is replaced by a warhead".[92] The Shavit launcher is a mobile one, which too confirms the military origins of the capability. Israel had commenced its acquisition of missiles after contracting a French firm, Dassault Aviation, to develop a 500 km range ballistic missile with a 750 kg payload.[93] Israel received 12 such Jericho 1 missiles by 1969, had built a production line and had plans to have a total of 24-30 missiles, including ten programmed for nuclear warheads by the end of 1970.[94] French cooperation halted and Israel commenced indigenous development which gave way to a two-stage, solid propellant Jericho 2 long range ballistic missile capable of delivering 1,000 kg payload to 1,300 km. It was the Jericho 2, first launched into the Mediterranean in 1986, around which was based the three-stage Shavit that launched Israel's first satellite.[95] Currently, in its inventory of ballistic missiles, Israel also has the 5,000 km range ICBM, the Jericho 3, a 30 ton, two-stage, solid fuel missile capable of a 500 kg warhead, including a nuclear one.[96]

Israel's Space Endeavours

The origins of the Israeli space programme can be traced to the 1960s when the country focussed more on the fields of research and education than planning for a long-term programme in space. In July 1961, a two-stage sounding rocket was fired that carried its meteorological payload up to a height of 80 km. This was intended to show Israel's technological superiority over the Egyptian rocketry.[97] The decade of the 1970s saw development of infrastructure for space science and exploration. A major change in the narrative of the Israeli space programme came about in 1979 when Israel and Egypt ended their 30-year-long war and signed the historic peace treaty. Article III of the treaty prohibited the two countries from flying combat aircraft and operating reconnaissance flights over the other's air space and over the Sinai peninsular region between the two. Israel now could not fly its reconnaissance aircraft over Sinai to make sure Egypt was not moving missiles into the demilitarised area. Gathering intelligence about the activities of its neighbours actually became the strategic driver for Israel's national security. Not wanting to rely on anyone else for intelligence, spy satellites was the solution found. "The Israeli space programme was a byproduct of the 1979 peace treaty with Egypt", according to a statement in a publication of the Israeli Ministry of Foreign Affairs titled *Israel at Sixty - From Modest*

Beginning to a Vibrant State: 1948-2008, p. 101.[98] Ironically, Israel's military was initially unenthusiastic about the start of the space programme as it felt the funding could have been put to better use. The Ofeq satellite programme commenced in 1982 and the Israel Space Agency (ISA) was established in September 1983, which, along with the Israeli industry and academia, took to overseeing and coordinating the entire gamut of space operations, taking five years to launch the country's first satellite, the Ofeq-1.

Concurrent with the development and fielding of the ballistic missiles, Israel has employed Jericho technology as a base for its indigenous Shavit space launchers. Shavit 1 led further to developing Shavit 2, a higher performance launcher that could place a 400 kg payload in LEO. Shavit 2 commenced its operations with the launching of the Ofeq-7, a 300 kg spy satellite in June 2007.[99] It is the Shavit series of space launch vehicles that has put into orbit the variety of satellites that Israel has deployed and currently has operational.

Geopolitical compulsions that impinge upon Israel's launching westwards have already been discussed above. To offset the disadvantage of launching reduced payloads, Israel was compelled to focus on developing small-sized (300-400 kg[100]) high technology satellites and sub-systems. Israel currently operates different constellations of satellites for different roles. In the role of Earth observation,[101] the EROS (Earth Resource Observation Satellites) operates in the Sun-synchronous LEO as a commercial venture providing very high resolution imagery to the international market. EROS-B launched on a Russian launcher in April 2006 in a Sun-synchronous orbit, provides 0.7 m panchromatic resolution while EROS-C is planned to provide in addition 2.8 m multi-spectral resolution. Within the definition of the Earth observation satellites, Israel also operates the Ofeq series of satellites which are actually reconnaissance satellites for the military,[102] the latest of which was the 300 kg imaging 'spy' satellite, the Ofeq 9, launched aboard Shavit 2 on June 22, 2010.[103] The Ofeq series of satellites with optical sensors provide sub-metre resolution imagery which is downloaded when the satellites fly over the ground control station. It was reportedly the Ofeq 7 spy satellite that played a big role in Israel's pre-dawn air attack on the Syrian al-Kibar nuclear facility in September 2007, by being diverted to fly over Syria to provide intelligence on the targeted facility.[104]

The TecSAR satellites go a step further by being equipped with SAR that is all weather, 'seeing' through fog and cloud, during both day and night. Commencing 2007 as a collaborative venture with the Northrop Grumman Corporation of the USA, the TecSAR capability integrates ISR missions with the high resolution SAR imagery.[105] A secret launch contract with Antrix Corporation, the business branch of the Indian Space Research Organisation (ISRO) in 2005, enabled the launch of the TecSAR/Polaris satellite aboard the Indian space-horse, PSLV (Polar Satellite Launch Vehicle) from the Satish Dhawan Space Centre, Sriharikota Range in India in January 2008.[106] Its onboard dynamically steerable antenna allows it to be redirected into direction of the flight path and its X band multi-mode SAR gives a one metre resolution, the imagery of which is streamed in real-time over a high speed data link to the ground control. This 360 kg satellite was reported to be Israel's most advanced spacecraft at that time.[107]

In the vibrant field of communications and broadcasting, Israel operates the geosynchronous Amos (Afro-Mediterranean Orbital System) satellites. Amos-1, launched in May 1996 aboard the French Ariane 4 launcher, was the first in the series of satellites that provides commercial viability for TV broadcasts and telephone communications for its customers. Amos-2 and Amos-3, both launched by Russian rockets, operate at the GEO slot of 4 degree W and provides value addition in the Ku-band over the geographical spread of Europe, the Middle East and across the Atlantic to the US east coast. Amos-5, launched in December 2011 aboard the Russian Proton M rocket, functions as a platform for remote transmission and a data hub for Internet providers, TV and telecommunications.[108] The US$ 200 million five-ton Amos-6 being built by Israel Aerospace Industries (IAI) in partnership with Canada and France is slated for launch in 2015 aboard SpaceX's Falcon launcher.

To further hone the technological achievements in the field of small sized satellites, Israel has worked towards further miniaturisation. One such venture came about by the ISA's cooperation with Technion, the Israeli Institute of Technology, to develop a constellation of three to four low-cost, low-powered microsats for Earth observation and communications. Techsat 1A's launch aboard the Russian Start ICBM in 1994 failed, but Techsat 1B, a digital communications satellite was successfully launched in 1998 by the Ukrainian Zenit rocket in a LEO. This 18-inch cube-shaped three-axis stabilised satellite, designed by students of Technion, has solar panels on all its sides and carries a miniaturised low-resolution camera, a UV spectro-radiometer and X-ray detector amongst others, all in a satellite weighing a mere 48 kg. While Technion slowed down its work on microsats due to lack of funding, ISA continues to pursue their development to establish itself as a competitive commercial provider. Israel has also contemplated use of nanosats as a hard-kill option for disabling satellites,[109] though inadequate funds appear to have slowed work on the same.

Israel Aerospace Industries (IAI), formerly Israel Aircraft Industries, a state-owned conglomerate, is Israel's largest aerospace and defence company, and also the largest exporter, with defence related exports to over 70 countries. It is the IAI that is the prime contractor for the three-stage Shavit 2 launch vehicle and the sole builder of Israel's military spy satellites.[110]

International Cooperation

Given the unique geostrategic environment that Israel finds itself in, it has had to rely on foreign technological support and collaboration right from its early years of security-enhancing capability, some of which led to activities towards space exploration. France and the USA have been rather active in their support – the former in missile technology and the latter in almost all activities in the military-techno arena, and especially post-1980s in space when Israeli scientists participated in numerous NASA projects. Israel had also fielded one of its fighter pilots on a Space Shuttle flight, unfortunately, the ill-fated Columbia that burnt up on reentry in February 2003. With Shavit 2's payload carrying limited to 400 kg and that too restricted to retrograde launches, Israel has taken to relying on France, Russia and India for space launch services. The USA has been involved in the TecSAR project, including purchase of radar imagery from these satellites. From 1996 onwards, Israel has been the only non-European country to be fully associated with the research funding programmes of the EU, with Israel contributing Euro 440 million to the Euro 50.5 billion for the programme spanning 2007-13.[111] With India, Israel signed an agreement in 2003 to launch its triple-telescope platform, the TAUVEX (Tel

Aviv University UV Explorer), aboard the Indian GSLV launch vehicle. With France, the SPOT project enabled Israel to set up a receiving station in Israel to receive SPOT imagery directly. The year 2005 saw ISA and France partnering for the Venus project, an electric propulsion microsat being designed for environment monitoring and Earth observation. Another agreement has been signed with Canada for cooperation in the field of micro- and minisats, hyper spectral satellites and space assisted education programmes for the youth.[112] Amongst the applications for the common man, one of the most commonly used is the 2D and 3D imagery of the Earth. Google Earth buys its images for its maps from four satellite companies, one of them being from Israel.[113]

Prospects for the Future

As fast paced, extensive and innovation driven progress in the arena of activities in space has been for Israel in the last two and half decades, equally ambitious are its plans for the coming decades. Predominant in its range of immediate activities will be the continuing focus on launching photoreconnaissance and communications satellites. The under development 400 kg Ofeq 10 satellite, the largest and the most capable of Israel's electro-optical imaging satellites, aims to provide a 50 cm resolution.[114] Eros C for the civil-use photography for the Israeli firm ImageSat and Amos 4 in the field of communications for the firm Spacecom, are two satellites currently under development of the six being planned. Others could well be military spy satellites and some for foreign customers.[115] The new space programme drafted by ISA and the Ministry of Science, based on the premise that Israel is losing its relative edge in the field of space due to its focus on military satellites, intends to correct the situation by aiming to get a significant slice of the global space market, currently estimated to reach a hundred billion dollars a year.[116] The new programme identifies the need to channelise military technologies towards the civilian market and as a result, envisions 70 percent of government grants up to 2016 devoted to Israeli industry for the development of space products for civilian uses. This will fructify in developments in software intensive digital communications for the new communications satellites, while the field of Earth observation and research will bank upon microsats.

In the field of space launch vehicles, in an attempt to break away from the constraints of westward launches that impinge upon the payload, Israel is considering the option of launching its Shavit 2 from an airborne platform. In the conceptual stage as yet, Israel is devising a way to launch its indigenous launcher from a modified Boeing 747 aircraft in international air space high above the Indian Ocean. This option would enable launching away from Israel's congested air space amidst hostile neighbours, from an altitude of about 12,000 m to the east, in the direction of the Earth's rotation, with the added benefit of almost doubling the lift capacity of the ground-launched Shavit 2.[117] In addition to the Boeing-boost option, Israel is considering other proposals for launch that include an entirely new rocket much heavier than the Shavit 2 and adapting Shavit 2 for sea-based launches in international waters.

For a small sized state with negligible strategic depth, in a hostile neighbourhood, Israel since its independence in 1948, relying on its resolve for sovereignty, has indeed stood its own. Building up its military-industrial infrastructure with extensive focus on R&D and innovation, it has aimed for, and rather commendably achieved, indigenisation in most of its activities towards national security. Having carved its name among prominent spacefaring countries of the world, Israel acknowledges the necessity of international cooperation. It has entered into agreements with almost all the nations that have a role to play in space, and in all that it cooperates and collaborates, Israel has always maintained its extremely zealous nationalist fervour for its sovereignty and security.

Iran Reaches Space

Iran joined the exclusive club of countries that have indigenous capability to launch their own satellites on February 2, 2009, by launching the Omid, a small communications satellite on its Safir 2 rocket into LEO. Iranian President Mahmoud Ahmadinejad ordered the launch, well timed to mark the 30th anniversary celebrations of the 1979 Islamic Revolution and remarked that the satellite was a "step toward justice and peace".[118] The Iranian narrative of launch-into-space too has flowed from its ballistic missile programme, a large part of whose development and progress had been kept concealed from the world and it was only the sporadic activities in rocket launches and an odd attempt or two to launch satellites aboard Russian rockets that gave an indication of the seriousness with which Iran pursued this vision. This singular national pride evoking achievement charts a special significance for Iran as it came about after 30 years of Western sanctions and international isolation.

Capability Building in Satellite Technology

Iran's interest in space matters actually dates back to the late 1950s when the UN established COPUOS (Committee on the Peaceful Uses of Outer Space) and Iran as one of the 24 founding fathers, went about pursuing its vision of using space and its technologies for peaceful purposes and for the country's welfare. One of the first fields of space applications that Iran addressed was that of communications, by joining Intelsat (International Telecommunications Satellite Organisation)[119] in 1970, a year after Iran had constructed two 30-m diameter antennae on its soil. Though the need for state owned satellite-based communications was apparent even in the 1970s, it was not till the 1990s that Iran put in place plans for the Zohreh satellite system – a series of geosynchronous satellites for which Iran had duly reserved three slots in GEO. India, France and Russia at various times were being considered for building these satellites for Iran.[120] The single satellite contract was finally awarded to Russia in 2005[121] and though not a part of this contract, was followed by the launch of Iran's first spy satellite, the Sina 1, aboard Russia's Kosmos 3M rocket into a Sun-synchronous orbit, enabling Iran to be the 43rd country to have its own satellite in orbit.[122] Unfortunately for the Iranians, in 2009, Russia refused to play along with Iran's space programme[123] and that made Iran announce that it would build and launch the Zohreh indigenously some time in 2014.

While the Zohreh project faced upheavals almost from the start, other projects, especially in the ground segment, did achieve more than moderate success. Besides Intelsat, Iran contracted with other commercial satellite providers like Eutelsat, HotBird and Telestar-5 that added additional transponders in the Iranian network for communications and TV services. In the field of meteorology, Iran acquired antennae and satellite receivers for meteorology satellite services from the USA, Russia, Europe and China.

Towards remote sensing activities, the USA's first cooperative venture with Iran in 1973 led to imageries from the satellite Landsat being received at Mahdasht Satellite Receiving Station (MSRS) operated by the Iranian Remote Sensing Centre. MSRS over the years has helped to identify areas prone to earthquakes, floods, other natural disasters; it monitors urban and remote areas and investigates greenhouse gas emissions, air pollution and the environment. MSRS has evolved into the Mahdasht Space Centre (MSC) and comprises well-established multi-task ground space complexes and laboratories.[124]

Iranian Space Agency

In the 1970s, when plans were afoot for the Zohreh project, several other Iranian organisations pursued plans to send small satellites into space, which could then set the pace for launching military intelligence-gathering satellites.[125] The plan for establishing a central agency, the Iranian Space Agency (ISA), to coordinate and monitor various organisations and agencies involved in space related activities, first arose in 1977 but was put on hold as Iran was hit by its revolution that brought about a regime change in 1979, followed by the prolonged war with Iraq (1980-88). Approved by the Iranian Parliament in 2003, ISA was finally established in 2004 with the brief to generate programmes for the peaceful applications of space technologies, with respect to satellite and launch vehicle programmes, as well as promotion of the ground segment.[126]

Notable amongst various agencies that contribute to the country's space programme is the Aerospace Research Institute (ARI) working in the area of fundamental and applied space research and the Iran Electronics Industry (IEI) that develops and builds the hardware. ARI is an academic organisation founded in 2000, affiliated to the Ministry of Science, Research and Technology, reportedly subordinate to the Ministry of Defence,[127] focussed in meeting research requirements in aerospace technologies at the national level and since 2007, providing a master's degree in space engineering.[128] The range of research covers the entire spectrum of space activity: sounding rockets, rocket engines, launch vehicles, reentry vehicles, satellites, payloads, orbital debris, space medicine and the commercialisation of space.

With eight subsidiaries, over 100 different kinds of electronic products and 5,200 experienced personnel (as of 2009), IEI is considered to be the largest electronic conglomerate in Iran[129] and a major producer of military products. Also called SAIran, IEI, founded in 1972, is a non-civilian wholly-owned subsidiary of the Ministry of Defence and Armed Forces Logistics[130] providing a wide range of customer support to its military[131], aviation and its aerospace wings. The subsidiary of IEI that deals exclusively in Iran's satellites is the Iran Space Industries Group.

Developing Satellites

Besides the Zohreh project, Iran announced its plan in 1999 to design and manufacture the Mesbah, a research satellite to be jointly developed initially with Russia and then Italy. Iran invested $ 10 million in its first indigenous-content 65 kg micro satellite project and sent the completed satellite to Italy in 2006 for final tests prior to sending it to Russia for the launch.[132] Italy deviated (*sic*) from handing over the satellite to Iran citing international sanctions in the wake of six resolutions by the UN Security Council against the perceived military nuclear programme of Iran.[133]

Well before the recent sanctions against Iran by the UN, Russia launched Iran's first satellite, the 160 kg Sina-1, officially dubbed a remote sensing and communications satellite into a Sun-synchronous polar orbit in October 2005. Built by the Russian Polyot design bureau for $ 15 million,[134] the satellite is equipped with four folding solar panels, communications payload and two cameras. One camera provides a 50 m resolution over a 50 km swath in panchromatic mode and the other offers 250 m in the multispectral mode, performance parameters not quite adequate for use in military intelligence gathering missions. Another cooperative venture that met with success was a satellite built jointly with Thailand and launched aboard the Chinese Long March 2C rocket in September 2008. The 490 kg Sun-synchronous satellite in LEO is aimed at disaster and environmental monitoring, civilian remote sensing and communications experiments.[135]

A major milestone in the Iran's space programme was achieved when success came Iran's way in the capability of launching a home-made satellite from an indigenous space launch vehicle, which was demonstrated to the world with the launch of the Omid aboard Iran's own Safir-2 rocket in February 2009. The cube shaped 27 kg communications microsat completed 1,300 orbits before burning up in the atmosphere over two months later, as per its stated lifespan. Iran could now boast of possessing technology not only in the field of micro satellites but, more importantly, in the capability of operationalising indigenous satellites and launch vehicles. With this capability firmly in place, Iran could now seek to increase the capacity of its satellite-carrier rockets to carry more weight.[136] Technology enhancements in the field of microsats continued, and progressed to launch a 15.3 kg image-collecting satellite, the Rasad, on its Safir 1A rocket on June 15, 2011, with a 260 km altitude orbit[137] beaming back pictures of 150 m resolution. This event in the Islamic Republic of Iran, yet again, evoked the world's immediate condemnation and severe criticism of the threat posed by Iran's development of ballistic missile technology and the nuclear or conventional warhead carrying capability of such rockets. Mark Fitzpatrick, a proliferation expert at the International Institute of Strategic Studies, London, looks at the issue pragmatically and counter-argues, "Iran could not easily adapt this space launcher for an intercontinental ballistic missile. A warhead would be several times heavier than this satellite, which is lighter than the first one they launched. The military significance of this can be easily exaggerated".[138]

In pursuit of its vision of being a respectable space player in the global arena, Iran has enlarged the scope of its activities in space. A new satellite launch vehicle, the Kavoshgar-3, took to the skies on February 3, 2010. The payload this time included a mouse, two turtles and some worms which was blasted off into sub-orbital space and parachuted back alive in an attempt to study the condition and movement of the animals in a mini-environmental laboratory[139] making Iran the sixth country in the world to do so. This research flight of 'animals in space' was found to be 'provocative' by the USA.[140] Iran-born US citizen Anousheh Ansari achieved the singular honour of being the first Muslim woman in space and the first woman in the world to pay for her own 'space fare' estimated to be $ 20 million, when she undertook a 10-day journey aboard the Russian Soyuz spacecraft to the ISS in September 2006.[141] Quite akin to other spacefaring countries, Iran, too, has involved its academia to actively participate in some parts of its space programme. Scientists and youth from the Iran University of Space and Technology partnered with ISA to build a 50 kg remote sensing satellite named Navid-e Elm-o Sanat. ISA reported on its official website on February 6, 2012, about its launcher-to-be 'Safir-e Navid' without giving the date of the launch[142] while the Iranian press dates the event of the launch to February 3, 2012.[143]

Iran's Rocketry that led to Space Launchers

It was the Safir 2 space launcher that placed the Omid in orbit in 2009. The rocket technology in the Safir is based on the technology found in the Shahab series of Iran's ballistic missiles, which, in turn, are derived from the North Korean No-Dong missile. The evolving saga of Iran's missiles traces its origins to the Iran-Iraq War (1980-88). Iraq's modified Scud missiles outnumbered and outranged those possessed by Iran. Iran sought, and North Korea obliged with, liquid-fuelled Scud-Bs, which ironically, were reverse-engineered from the Scuds that North Korea had acquired from Egypt in the early 1980s.[144] The Egyptian Scuds turned into the North Korean No-Dongs, came to Iran as the Scud-Bs and post-war Iran based this technology to produce its Shahabs. The Iranian version of the Scud-B became the Shahab 1 while its successor, the Shahab 2, a variant of the Scud-C, iterated to the Shahab 3[145] which is a military single-stage liquid-fuelled road-mobile medium-range ballistic missile with a range up to 1,200 km which was test flown in 1998. Some variants of this missile are claimed to have achieved a 2,000 km range.

The two-stage Safir launch vehicle was first test-fired in August 2008, without a satellite payload. Televised footage of this launch and the successful Omid-bearing launch in February 2009 indicated its liquid-fuelled first stage to be that of the Shahab 3, derived from the Scud/No-Dong technology while the second-stage, also liquid-fuelled, seemed to be indigenous. The Safir 2 is designed to place a payload of up to 25 kg into LEO at 245 km altitude,[146] while some reports suggest 27 kg into 250-500 km LEO orbit.[147] It has a lighter variant in the form of the Kavoshgar, designed to be a civilian sub-orbital sounding rocket for research related payloads.

The Years Ahead

Satellites that would address Iran's needs in the years ahead will necessarily have to be heavier than the ones that Iran has orbited in the last three years. Iran's current handicap is not the capability to develop heavier satellites but is defined by the limited-payload launch capability of its Safir space launchers. Indigenous Safirs orbited the Omid (27 kg), Navid-e Elm-o Sanat (50 kg) and Rasad-1 (15.3 kg) but for the heavier Sina-1 (160 kg), the tri-nation 490 kg Earth imaging satellite, Environment 1, launched in September 2008, and the unsuccessful Mesbah (65 kg), Iran had to bank upon the foreign launch platforms of Russia, China and Russia respectively. Thus, for Iran to see any fruitful progress in its space programme, developing more powerful space launchers looms as a strategic imperative.

There is no denying that Iran's civil-space origins are attributed to the country's clandestine ballistic missile pursuits. As missile ranges improved over the years, so did the capability of space launch rockets. Developmental work is currently ongoing for a heavier lift launch vehicle. One such project entails development of the Safir Block II that adds two solid motor strap-on boosters to the first stage of the Safir 2 and possibly a new solid-fuelled third stage, enabling this variant to either double the payload capacity or increase the orbital altitude to 700 km.[148] Another developmental activity aims at a new generation satellite carrier, the Simorgh that was unveiled on February 3, 2010. The launch vehicle employs a new generation of two-stage liquid-fuelled engines in the Simorgh 3 that could enable the launcher to place a 60-100 kg payload into a 500 km LEO.[149] Further refinements in replacing its second stage with the second stage of North Korea's Unha 2/No-Dong-B missile or Taep'o-Dong-2B booster could give the launcher a capability of placing 700 kg in a 1,000 km orbit[150] which then could be converted to an ICBM[151] if it was to be in Iran's strategic and domestic interest. The above payload carrying capability is also listed on ISA's website.[152]

Armed with this powerful launch capability, Iran then has plans for a whole series of new generation of capability-enhancing satellites. Three of these that are listed on the ISA website are the Tolou, Mesbah 2 and Navid, the first in the role of remote sensing and the latter two for communications, all aboard the Simorgh.[153] The Tolou is reportedly an 80 kg satellite for a 400 km orbit that would provide 50 m resolution imagery[154] while IEI claims to have made a camera for fitting on the Tolou that will provide a 20 m resolution.[155] The Fajr is another remote sensing satellite but of 50 kg weight for 250-400 km altitude elliptical orbit with a 500-1,000 m resolution that will also carry a locally made GPS system.

For telecommunications, the Mesbah-2 is planned to be a microsat in LEO, operating in the amateur radio frequency band with the aim of research, e-mail and store-and-forward data communications.[156] Iran has plans for the launch of a 1,800 kg satellite, the Qaem, some time in 2017 and with a life of 15 years, the Qaem is aimed to meet Iran's telecommunications, radio and TV needs of the future.[157] In the field of space situation awareness, Iran plans on launching the Nasir 1, an indigenously designed 50 kg microsat for precise location of other satellites in orbit.[158]

In the realm of international cooperation, Iran has entered into an agreement with member countries of the Asia-Pacific Space Cooperation Organisation (APSCO) for developing and launching 10 satellite projects in the 500-600 kg class for research, remote sensing and telecommunications.[159] A cooperative approach is being taken in the Besharat satellite project too. Iran will launch this remote sensing satellite that is being built by Iran in collaboration with some Muslim member countries of the Organisation of Islamic Conference.[160] The length and breadth of Iran's ambitions in space can also be gauged by its plans for a manned space mission by the year 2018.[161]

Firm in its ideology of self-reliance and belief in its sovereign right to space technologies, Iran has continued to address space-based activity as an essential ingredient for its techno-nationalistic honour and prestige. While Pakistan has staked the claim of being the first nuclear state amongst the Islamic countries, Iran has achieved the status of being the first Islamic country to launch its own satellite and will continue to maintain that edge. It is with this single-minded pursuit that Iran has planned to achieve the *numero uno* status in the Middle East region in the field of applications in space technology, by the year 2025.[162]

North Korea Eyes a Slice of the Space-Pie

"North Korea rocket launch fails" read the title of the news report in BBC News on April 13, 2012[163] stating that the launch came amidst international warnings against North Korea's nuclear programme and the event prompted immediate global condemnation. The news report quoted the state-run (North) Korean Central News Agency's statement that it was a North Korean Earth observation satellite, the Kwangmyongsong-3 (KMS-3) that was launched by the three-stage carrier-rocket Unha-3, but had failed to enter its preset orbit. The US stated that the launch constituted a disguised test of a long range missile technology banned under UN resolutions.

This was not the first time that a satellite-launch bluff was raised by North Korea. The Democratic People's Republic of Korea (DPRK) claimed to have launched its first satellite, the KMS-1 experimental satellite aboard the Taepodong-1 (TPD-1) way back in August 1998 too, though the spacecraft tracking network in the US Space Command was unable to confirm the presence of this satellite in orbit.[164] This was followed by yet another claim of launching a satellite in April 2009, that of the KMS-2, aboard the Unha-2, which too was found to be 'absent' in space by the tracking networks of the USA and Russia. What was common in all the three cases was the DPRK's hidden agenda and the secrecy behind the country's space related motives and activities, leading the USA and most of its allies to regard the events as a North Korean subterfuge to disguise its actual activity of technology enhancements in ballistic missile capability. What was unique to the satellite-launch of 2012 was the North Korean state agency admitting to a failure, as against the first two that had been declared successes.

It was finally on December 12, 2012, that North Korea's ambition as a spacefarer bore fruit after 15 years. A replacement satellite of the KMS 3 was launched aboard the Unha-3 amidst scepticism and trepidation in the Western world, viewing the launch as yet another missile test by a rogue nation. The suspicion was, however, laid to rest as US space tracking stations confirmed within three hours of the launch that it indeed was a satellite that had reached orbit.[165]

Missiles that Aimed to be Space Launchers

The Soviet-backed DPRK crossed the 38th parallel and attacked the pro-Western Republic of Korea (ROK) in 1950 in an attempt to reunify Korea. The three-year Korean War proved to be a failed attempt by North Korea in its aim of taking South Korea by force and since then the military-oriented DPRK has made every endeavour to be techno-militarily stronger than its arch rival in the South. In 1966, the DPRK put in place funds and resources for the procurement and production of ballistic missiles. It appeared only natural that the DPRK would look towards the USSR and China to bolster its missile arsenal and it commenced receiving technical assistance from both. Scud rockets from the USSR and an agreement with China in the 1970s for joint development of liquid-fuelled ballistic missiles led to the North Korean Nodong (ND) missile a decade later.[166] It was the North Korean Scud variant which later found customers in Iran, Syria, Egypt and Pakistan, under the names of ND and Hwasong.[167]

There is another report on the missile's origins in North Korea, which suggests that some tactical missiles, the Frog 5 and Frog 7A,[168] came from the USSR in 1969 while the first Scuds came via Egypt in 1976, and by 1984, North Korea was building its own Scud-Bs of 300 km range and had developed new missiles, the 500 km range Scud-C and a medium-range missile, the ND, with an estimated range of 1,300 km, with Scud and ND technologies combined to give a long-range two-stage missile, the 2,900 km range Taepodong (TPD) or Paektusan-1.[169] The TPD-1's test flight in August 1998 was the launch that the DPRK claimed four days later, of its first satellite, the KMS-1. It was this very launch whose trajectory flew over Japan and galvanised the shift in Japan's space strategy from a purely peaceful one to one that initiated development of its first intelligence satellite. It was this very launch that aroused the security concerns in South Korea, prompting it to enhance its missile range to 300 km and enter the Missile Technology Control Regime (MTCR), allowing it access to superior space technology from other member states.

There has been much cooperative inter-play between North Korea and Iran in the development of their respective ballistic missiles and space launchers. North Korea's TPD-1 evolved into the two-stage TPD-2 with a range up to 10,000 km. It was this TPD-2 modified to a three-stage rocket-carrier, the Unha-2 that attempted but failed to launch the second satellite, the KMS-2, in April 2009. In the field of ballistic missiles, North Korea now is replacing the TPD-2 with a newer two- or three-stage missile with a range of 15,000 km.[170] Towards space launch capability, research and development is currently on to develop the TPD-3, dedicated for launching geostationary satellites, with space ambitions going further towards a manned space mission planned for in 2016/2017.[171]

Ballistic missiles being customised as space launchers would not have been new to the world order as the precedent had already been established by most, if not all launch-capable countries. This corollary in the case of North Korea had been initiated some time in the mid-1980s with the establishment of the state-controlled space agency, the Korean Committee of Space Technology (KCST)[172] under the country's military control, which was further given a fillip in April 1993 with approval accorded for development of a spacecraft. Despite UN sanctions and the US' threat of embargo on aid, the DPRK has entrusted KCST with the ongoing operations of its launch facility, the Unha space launcher and the KMS satellites.

Motives Behind DPRK's Space Ambitions

North Korea's aim to be techno-militarily superior and maintain that lead over South Korea relied on a singular fundamental premise, that of 'military first'. It was this principle that governed budget allocations, the majority of which were channelised to its military, largely concentrating on its nuclear and missile programmes. A major shift in its strategy of ideological influence over South Korea to one of regime survival came about by, firstly, North Korea's loss of military and financial aid from the USSR after the latter's break-up in 1991 and, secondly, by South Korea's burgeoning technology-driven economy.[173] The DPRK's continuing with space launches to disguise its development activity of the UN-sanctioned missile programme may not necessarily be a far-fetched idea, viewed in the context that the missile programme met with opposition from both its ally, China, and friendly neighbour, Russia. The extreme infrequency of the DPRK missile tests and the three unsuccessful attempts at satellite launches suggest that the DPRK, in fact, has no sophisticated or devoted satellite programme, nor any serious plans to develop such an industry. Future rocket tests in all likelihood would be an attempt to steal publicity from South Korea's space achievements. Notwithstanding the demonstrated indigenous launch of a satellite in 2012, North Korea is unlikely to emerge as a space player of any major significance in the times to come, seemingly content for the time being with collaborative ventures with countries in the Middle East and South Asia in the fields of missile technology and construction services.[174]

South Korea Takes to Space

Entry into Space

It was January 30, 2013, that the Republic of Korea (ROK), or South Korea, realised its ambition of orbiting its satellite aboard an indigenous space launcher. The 140 ton Korea Space Launch Vehicle-1 (KSLV-1) launched the 100 kg Science and Technology Satellite-2C (STSat-2C) into a LEO, from ROK's soil after two previous failed attempts in 2009 and 2010 and within seven weeks of its arch rival, North Korea, achieving the same status in December 2012, making South Korea the 11th country in the world to demonstrate its indigenous space launch capability. The ROK's space programme has been relatively a late starter amongst the spacefaring nations though the country had by then already orbited almost a dozen satellites through foreign launchers.

In 1992, South Korea became the 22nd country to have a satellite in orbit. The research satellite KITsat-1, took to the skies on August 10, 1992, aboard Ariane-4, from the Guyana Space Centre. The 48.6 kg satellite was designed by the Surrey Satellite Technology Ltd, England, and under technology transfer, built by post-graduate engineering students of the Korea Advanced Institute of Science and Technology (KAIST). A foreign designed satellite launched from a foreign space launcher might not amount to much but for the fact that South Korea's entry into this field was so recent that it had begun its space programme only in the late 1980s. Driven by the domestic need for a strong technology development base, the ROK in 1985 issued a ten-year scientific-technical plan that called for a series of specific space related projects.[175] Establishment of the Korean Aerospace Research Institute (KARI) followed in October 1989, to create a competitive air and space industry within the country and the ROK took to working with European and American companies to gain knowhow about satellite building and related technologies. This led to the ROK's first satellite, the Kitsat-1, in August 1992 to orbit.

Attempts at Indigenising Space Launchers

While the space endeavour might have been a late-starter, the ROK's experience with rockets and missiles dates back to the 1970s when the emerging North Korean missile activities forced the ROK to initiate a surface-to-surface missile programme in 1971. Technology, however, remained an issue with the missile programme. Fearing further destabilising flashpoints in the already volatile neighbourhood, the USA in 1979, impressed upon the ROK to limit its future missile's range to 180 km in exchange of receiving US assistance in missiles and related technologies: 180 km was enough to bring within range the North Korean forces on one of the most heavily fortified borders in the world, but not enough to take on North Korea's capital. This limitation in essence tethered South Korean missile ambitions to short range defence oriented objectives only.[176] By the 1990s, Korea had developed an independent capability to manufacture solid propellant rocket motors of up to one tonne mass and followed it up with a new Korean-made monolithic booster stage which together with a modified guidance system, produced a surface-to-surface missile of around 400 km range.[177] Within the limitation of range imposed on it, South Korea commenced work on sounding rockets for scientific experiments. In 1990, KARI was funded to build the first indigenous sounding rocket, the Korean Sounding Rocket (KSR), using a single stage solid-fuelled motor. The KSR-1 flew in 1993 to a height of 75 km and the two-stage version, the KSR-2 flew in 1997 and 1998 to an altitude of 150 km. To switch to the realm of civilian space-launch, the ROK had to move from the solid-fuelled military missile to the more controllable liquid-fuelled rocket. This need led to planning for a liquid-fuelled KSR-3 as the first step towards its ambitions in space.

Launch Capability Enhancement

Events that unfolded in the late 1990s in the region changed the trajectory of South Korea's growth in space. North Korea's test-firing its Taepodong-1 missile with an estimated range of more than 1,300 km and an attempted satellite launch in August 1998, prompted South Korea in 1999 to ask Washington to agree to extend South Korea's missile range to 300 km for deployment and to 500 km for scientific research and development, which was agreed to in principle by the two governments.[178] An add-on to this agreement, in fact, a bigger benefit, was the ROK joining the Missile Technology Control Regime (MTCR) in March 2001 that allowed the country to gain access to rocket technology from other MTCR member countries. The ROK government promptly announced a $ 4.26 billion investment in its space programme[179] and ever since, has pursued its indigenous space programme focussed on the following three key-objectives:[180]

- Indigenous launch capability of microsats by 2005.
- Indigenous development and launch of LEO multi-purpose satellites by 2010.
- Being among the top ten countries in the space industry by 2015.

The liquid-fuelled KSR-3 was finally launched in November 2002 to a height of 42 km over a distance of 84 km[181] but further development was shelved as by now, new avenues from MTCR members offered more attractive superior-technology alternatives for a space launcher. South Korea decided to leap-frog the indigenous development by contracting with foreign partners, looking first towards the US, which did not oblige. The ROK signed a US$ 200 million pact with Russia in 2004 for the transfer of the Russian under-development Angara-derived booster for developing the Korean Space Launch Vehicle (KSLV), and cooperation in the construction of South Korea's launch facility at the Naro Space Centre.[182] The KSLV-1 was designed around the Russian first-stage that was mated to the South Korean solid-fuelled second-stage and was launched with the South Korean satellite STSat-2 in August 2009. The two stages operated successfully but the satellite failed to release and fell back into the atmosphere. The second KSLV-1 launch in June 2010 too ended in failure and it was the third attempt in 2013 that brought success and the concomitant surge in national pride, more so as North Korea had already achieved that success just seven weeks earlier.

Satellite Technologies

Almost parallel to events taking shape in the ROK's attempts at developing indigenous space launch capability, work on designing and building satellites received equal if not more attention. One study comparing South Korea's position with respect to China's in 13 key technologies, quoted in RAND Corporation's analysis in 2005, found "Korea is relatively strong in IT-related areas but weaker in aerospace", while the RAND Corporation estimated that South Korea lagged behind China by about four years overall in terms of space technology, though it fared better in commercial applications.[183] China's ASAT test in 2007 and the world's understanding of China's true potential thereafter, in all likelihood, has made China's space capability larger than what was perceived in 2005 and further widened the gap between the two, but that notwithstanding, it does not take away the credit from South Korea's achievements.

KITsat alias Uribyol: The KITsat-1's launch in August 1992 was followed by a near identical and purely indigenous KITsat-2 in 1993 by Ariane 4, carrying Earth imaging and digital store-and-forward experimental equipment essentially to promote the domestic microelectronics industry. The 110 kg remote sensing KITsat-3 was launched in 1999 aboard the Indian Polar Satellite Launch Vehicle (PSLV) into a Sun-synchronous orbit and provided 13.5 m resolution multi-spectral imagery, paving the way for engineering small satellites for Earth observation. In continuation of this series, the KITsat-4, also called STsat-1, designed to be the ROK's first satellite dedicated to the national space science programme, was launched by Russia's Cosmos 3M rocket in 2003.[184]

KoreaSat alias Mugunghwa: In the field of commercial broadcasting and communications, South Korea has deployed the KoreaSat series of geosynchronous satellites, commencing 1995 with the aim of putting in place a satellite communications network for domestic and regional purposes, under the aegis of a private corporation giant, the Korean Telecommunication

(KT) Corporation. KoreaSat-1 to KoreaSat-3 were funded by KT and launched by foreign contractors. This system underwent a change when the ROK's Ministry of Information and Communications (MoI&C) unveiled plans for its first military satellite.[185] The MoI&C, KT and the MoD jointly funded KoreaSat-5[186], a dual-purpose radar satellite and South Korea's first satellite to have a stated military mission. The 4.5 ton KoreaSat-5, built and launched by France in 2006 into a GEO, carries 24 transponders in the Ku band for commercial purposes and exclusive for the military, four in the Ka and eight in the SHF bands.[187] The 2.6 ton KoreaSat-6 followed in 2010, again built and launched by France, orbiting in GEO as a purely commercial communications satellite with 30 onboard Ku-band transponders.[188]

KOMPsat alias Arirang: This series of satellites currently operates three multi-purpose Sun-synchronous satellites essentially for remote sensing and communications. KOMPsat-1, built by KARI with US assistance for design and launch, began operations in 1999, carrying a payload for 10 m resolution imagery and a low-resolution camera for ocean and Earth resource monitoring. The KOMPsat-2, launched in 2006 from Russia with enhanced capabilities, is equipped with a high-powered digital multi-spectral camera that provides one metre resolution panchromatic imagery for mapping the Korean Peninsula. The satellite`s main mission is to provide geographical information system-based data of the Korean Peninsula. It covers information such as hydrology, road networks, urban mapping and demographic data. The multipurpose satellite beams information on changing natural environments to help provide preventive measures against natural disasters. It scans disaster-stricken areas and surveys the country`s agriculture, fishery and forestry resources.[189] The fact is that KOMPsat-2 is a dual use civil/military satellite with obvious reconnaissance capability that the one metre resolution imagery can provide.[190] The KOMPsat-3, launched in May 2012, aboard the Japanese H2A rocket, has an array of infra-red sensors and an optical camera with 0.7 m resolution, making South Korea one of the select few countries with sub-metre resolution capability.[191] South Korea intends to take this network further with plans to launch dual-use SAR satellites KOMPsat-5 and KOMPsat-6 in 2015 and an optical KOMPsat-7 in 2017.[192]

South Korea's Space Compulsions

With the disadvantage of being a late starter in its space programme, South Korea had little choice but to base its approach on a fast-track mode of development. While doing so, the ROK has perhaps exhibited the most politically 'balanced' strategy amongst the 'third wave' spacefarers. Coming out of its long dependency on the US, South Korea has sought to maintain its technology ties with a host of countries with the dual aim of propping up its own technology and preventing its possible isolation.[193] The lynchpin of its space programme lay in the imbalance of missile development in the two Koreas. While South Korea was tied down with the US mandate of missile ranges of up to 180 km and at the turn of the century to 300 km, North Korea in the late 1990s already had the Taepodong-1 with a range of 2,900 km in its inventory. It was this inequity in the national security that led South Korea to focus on satellites and space related research. The surge in space activities thereafter was such that by 2009, the country had reached 80 percent capability vis-a-vis the USA and Russia in the medium and low altitude satellite fields while the gap stands at 40 percent in geosynchronous and communications satellites.[194] One field of the space sector that has arrived on South Korean shores and is being capitalised upon is that of commercial space. While the ROK has just about entered the space-launcher league and the global satellite market remains highly competitive, the country seems to be focussing on a different segment, particularly those cash-strapped countries that may be attracted to its lower-priced space-derived products.[195] The ROK's entry into the select group of spacefarers now provides the country ample opportunities for broadening its horizons and in due course, addressing the ever burgeoning commercial space-launch sector.

Prospects for the Future

No other space project of the future defines South Korea's ambitions to carve a name for itself in the global field as much as the intrinsic desire to enhance its indigenous capability in the space-launch segment. The space launcher planned for the future will be the indigenous two- or three-stage liquid-fuelled KSLV-2 (or Naro-2) intended to place a one ton payload into a 300 km LEO.[196] KARI plans to complete this indigenous rocket by 2019[197] and then plans to offer commercial launch services by 2020.[198]

To broaden the scope of its space activities, South Korea, with Russian assistance, fielded the first Korean in space. The woman astronaut Yi So-yeon spent nine days aboard the ISS, travelling to and from the Earth on the Russian Soyuz rocket in April 2008. In the arena of beyond-Earth activity, South Korea in 2007 revealed its plans to send a lunar orbiter by 2021 and a lunar probe five years later. In the years ahead, South Korea, duly cognisant of space technology being a national strategic stepping stone, aims to be a major player in the business of space commerce in satellites and associated services. Towards this end, South Korea is likely to continue engaging with its international partners in areas where it has, as yet, not achieved the degree of technological maturity that it desired, quite like the way that the ROK looked to ESA, Russia and India for space-launches and partnered with the USA for its KOMPsat series of satellites. Collaborative space relations in the future with Russia, China, India, ESA and Japan, though at the risk of maginalising the USA from the ROK's projects,[199] would generate additional options for South Korea in aligning itself with nations that will shape the geopolitical world order in the years to come. South Korea of today appears well entrenched in its space strategy based on its domestic technical expertise and international cooperation, heading steadily towards its stated objective of becoming one among the top-ten space powers.

Satellite Capable but Launch-Capability Deficient

The nations that possess indigenous capability to launch their own satellites into space also provide launch services commercially to those countries that do not as yet possess the same. While scores of 'third-wave' countries have built their own satellites either indigenously or in cooperative ventures with others or in some cases, by outright purchase, for reasons of economics and of self-reliance, they too are attempting to achieve indigenous launch capability. Countries like Australia and Brazil, amongst others, have a rather well developed technological infrastructure in place for developing their own satellites, but the capability of launching them into orbit as yet has not reached the demonstrable stage. Many of these countries have been open to international cooperation and collaboration while some have closeted their efforts from others and most, fortunately, do endorse the stand against deploying weapons in space.

Indonesia's Budding Space Programme

The fourth most populated and resource rich country in the world is made up of some 17,500 islands, 6,000 of which are inhabited. The need for non-terrestrial communications for the widely dispersed islands and monitoring the vast maritime domain without committing a large fleet of surveillance ships and aircraft, led Indonesia to look towards space technologies as a solution for its needs. It established a National Institute for Aeronautics and Space (LAPAN, in Indonesian) as early as in 1963 and commenced its space programme in 1975 with the purchase of a communications satellite from the USA's Boeing. The Palapa-A1 was launched in 1976 followed by the Palapa-A2 in 1977, both launched by US Delta rockets. Since then, Indonesia has a geostationary satellite based broadcasting network largely aided by the USA's commercial industry[200] where the Palapa-As, having outlived their operational lives were replaced by the second generation US Hughes-built four Palapa-Bs and even bigger and more powerful Palapa-Cs thereafter. The latest in this series is the 4.1 ton Palapa-D, built by the French-Italian Thales Alenia consortium and launched by China's Long March 3B rocket in 2009. Kitted with 24 standard C-band, 11 extended C-Band and 5 Ku-band transponders, the Palapa-D provides telecommunication services to Indonesia, the Association of Southeast Asian Nations (ASEAN) countries, Asian countries, the Middle East and Australia.[201] Global players were not the only ones to engage with Indonesia in the field of satellites for communications, as the network for mobile cellular services for Indonesia was built under a regional cooperative venture with Thailand and Philippines.[202]

In the interests of its maritime domain, Indonesia's work on Earth observation commenced with receiving imagery from the US' Landsat in 1981 and then from the French SPOT in the mid-1990s. In the segment of microsats, Indonesia partnered with Germany's Technical University of Berlin to build its first remote sensing microsat, the 56 kg LAPAN Tubsat or LAPAN-A1, which some sources claim to be the country's first surveillance satellite.[203] Launched in January 2007 aboard the Indian Space Research Organisation's (ISRO's) PSLV C-7 rocket into a polar orbit, the Tubsat provides 5 m resolution imagery and 200 m resolution video imagery. Wanting to reduce the costs involved and to achieve a certain degree of self-reliance, LAPAN has commenced providing more and more indigenous content to its ongoing remote sensing satellite programme. Iterative efforts in this series of satellites are planned to be continued with LAPAN-A2 and LAPAN-A3, both to be integrated in Indonesia and launched by Indian rockets. Both the remote-sensing satellites, LAPAN-A2 and -A3, besides monitoring maritime traffic, in addition will carry payloads to support amateur radio communications.

Indonesia's rather modest achievements in satellite applications in support of the country's communications, remote sensing and weather monitoring has actually outstripped the success achieved in the field of launchers. The RPS (Indonesian, for Orbital Satellite Rocket) programme uses the classification of RX (Roket Eksperimental) followed by numbers that indicate the diameter of the rocket in millimetres. Commencing rocketry with the 30 kg solid-fuelled RX 100 that went up to 7 km altitude, LAPAN proceeded through RX 250 that achieved 70 km[204] altitude in 1987, RX 320 in 2008, and is currently toying with a three-stage experimental RX 420 which first flew in 2009 to a height of 53 km.[205] The 6.2 m tall rocket with a diameter of 42 cm and weighing 300 kg, RX 420 is the biggest rocket launched by Indonesia. Further plans for iterations on RX 420 at best could deploy only a 5 kg nanosat.[206]

Indonesia has announced plans for its first launch of a small satellite into a LEO in 2014 by its indigenous four-stage RPS based upon three stages of the RX 420 and one of the RX 320.[207] The rockets in Indonesian's current inventory have employed solid-fuelled stages, an issue that needs serious reconsideration and additional funding for the country to move into the more advantageous realm of liquid fuelled rockets. The additional funding would need to be a major departure from the current near-static budget allocations, to support further activities in its quest for self-reliance in placing a satellite in orbit, as planned for in 2014.

Malaysia's Space Debut

Malaysia joined the club of 'third wave' countries on January 12, 1996, when the Malaysia-East Asia Satellite-1 (Measat-1) achieved its intended geostationary orbit. The US Hughes-built telecommunications satellites Measat-1 and thereafter Measat-2 that followed ten months later, were launched by Arianne as a step towards establishing a communications network of geostationary satellites providing coverage as far as Eastern Australia. The Boeing-built 4.9 ton Measat-3 launched in 2006 aboard the Russian Proton rocket aimed to expand the country's communications capabilities[208] by its payload that carried 24 transponders each in the C- and Ku-bands[209] and was followed by the 2.3 ton Measat-3A aboard the Ukrainian Zenit rocket in 2009.[210]

Despite Malaysia's status as a newly developing industrialised market economy that boasts of the 30th largest Gross Domestic Product (GDP) in the world,[211] Malaysia has not appeared to be too favourable towards developing an indigenous technology base aimed at making it space-capable, but instead has chosen the collaborative approach with other space players, both in the fields of designing and building satellites and in launching them into orbit. Its second collaborative foray in the field of satellites was the joint venture with Britain's Surrey Satellite, Ltd., in developing remote-sensing 50 kg microsat TiungSat-1 that reached LEO aboard the Russian Dnepr rocket in 2000. The payload also included a meteorological Earth imaging camera and digital store-and-forward communications package. Its second remote-sensing satellite, the RazakSat, built in partnership with South Korea, was launched by the US' private enterprise SpaceX's Falcon 1 rocket in 2009, making it the first successful satellite-launch for the Falcon 1.[212]

Having seen the benefits of space technologies during the early years and duly noting the needed goal of inspiring the younger generation in careers in space, the Malaysian National Space Agency (MNSA), or Angkasa in Malay, was created in 2002.[213] The word *angkasa* is derived from the word *angkasawan*, the Malay word for astronaut. Was 'a man in space' also an agenda for MNSA? The subsequent event proved it so. A deal inked in 2005 between Russia and Malaysia gave the Malaysian Air Force 18 Sukhoi-30 MKM fighter aircraft and an offset clause[214] of the agreement between the two governments provided for Russia to train two Malaysians in space travel for 18 months and send one to the ISS. This led to 35-year-old orthopaedic surgeon, Dr. S. M. Shukor, being launched aboard the Russian Soyuz, spending ten days on the ISS in October 2007.

The past few years have seen Malaysia look towards developing more and more indigenous content in the satellites that were built under partnership with other countries. To be sufficiently self-reliant in this field, the Malaysian government aiming at expanding its satellite development allocated US$ 62 million for a five-year project for the remote-sensing RazakSat-2 satellite. The RazakSat-2 programme would be based on home-grown innovation and on its completion by 2015, would provide imagery for diverse purposes like mapping, agriculture, fishery as well as urban and coastal monitoring.[215] In its quest to rank amongst the globally recognised developed countries, Malaysia would need to look at the broader picture of committing itself to capability building and enhancements in the realm of space technology. Towards this end, Malaysia needs to exploit its close-to-Equator geographical location, which in turn throws up options for its geo-strategic and economic interests. Ideal, of course, would be an indigenous space-launcher capability, launching from its own soil close to the Equator. As space launch capability takes an enormous amount of time to hone and develop, Malaysia, in the interim, could build an independent modern launch facility on its soil and then offer the same commercially to its international partners for them to transport their launchers and launch them from Malaysia – an issue that finds favour due to the country's near-Equator location. For the short term, the country could aim to develop a totally indigenous satellite at the earliest and then explore the option of opening up this capability to international commerce.

Pakistan's Space Card

Pakistan's ambition to have a satellite in space bore fruit in July 1990 when the 52 kg microsat Badr 1, was launched aboard the Chinese Long March 2E rocket. The digital communications experimental satellite, Badr 1, carried a simple store and forward transponder and was built by the state agency, the Space and Upper Atmosphere Research Commission (SUPARCO). While contact was lost with the satellite a month later, the technological achievement boosted national pride and heralded Pakistan's entry into the 'third wave' spacefaring countries in the world.

While Pakistan enjoys the status of being the first Islamic country to have a running nuclear weapons programme and is one of the few countries in the world to possess ballistic missiles, its capability in the dimension of space has not been as advanced. It did indeed realise the advantages of a space programme as early as in 1961 when it set up SUPARCO, with the stated aim of launching indigenous satellites aboard indigenous launchers.[216]

Pakistan's Rocketry

The early 1960s was the era when the USA's space agency, NASA, helped Pakistan in sounding rockets, launching the first US rocket in June 1962 from Sonmiani Beach, 50 km west of Karachi. The rocket was a combination of two US rocket motors, the Nike and Cajun, which reached an altitude of 130 km, leading NASA to announce that the launch was the beginning of "a programme of continuing cooperation in space research of mutual interest". Cooperation was also extended to the extent that Pakistani scientists were trained at NASA space science centres. France pitched in with technology transfer for developing sounding rockets, German firms assisted in space research and supplied vital ingredients of solid rocket fuel and Great Britain helped with rocket launches.[217]

It was in the 1980s that the charter of SUPARCO, while seemingly civilian-oriented, swung towards pursuing military objectives. In 1981, SUPARCO announced plans to test a launcher in 1986 and by the mid-1980s, had established its own solid-fuelled rocket production facility replete with related test and tracking infrastructure. These facilities developed the tactical ballistic missiles of the 1980s and early-1990s, the 70-100 km range Hatf-1 and the 300 km range Hatf-2, based on designs and technology of the French sounding rocket, the Dauphin.[218] A turning point in the narrative of Pakistan's growth in missiles and launchers came about in June 1991 when the USA imposed sanctions on SUPARCO[219] and two other Chinese agencies for their activities related to missile proliferation. Now came the departure in Pakistan's policy in seeking foreign assistance – Pakistan's focus shifted from the USA and European partners, and came to seek assistance from North Korea and its all-weather ally, China. To move beyond the 300 km range Hatf-2, Pakistan desperately needed composites, specially alloys and Ammonium Perchlorate (AP), an oxidiser for solid propellants. In the year 1996 alone, three shiploads of AP totalling 320 tons, originating from North Korea and shipped through China, were seized by Taiwan and Hong Kong.[220] This seizure gives a fair idea of what did not get intercepted or seized. In July

1997, Pakistan tested the Hatf-3/Ghaznavi, a 400 km range missile based on China's nuclear-capable M-11 ballistic missile.[221] The move of China's M 11 missile components into Pakistan was being tracked by the US' Central Intelligence Agency (CIA) that concluded that China was not only selling missiles but also helping build a factory in Pakistan to manufacture them.[222] The Hatf series of ballistic missiles, thereafter progressively grew in range and capability, invariably with Chinese and North Korean assistance. The Hatf series also includes a 500 km nuclear capable cruise missile, Babur, which resulted from a technology transfer agreement with China.[223]

Pakistan's ballistic missile Shaheen II has achieved ranges of 3,500 km but the capability as yet has not been able to translate into any successful Space-Launch Vehicle (SLV). Dr. A. Q. Khan, the infamous father of Pakistan's nuclear programme, as Science Advisor to Pakistan's President, announced in 2001 that the country's first SLV had been entrusted to SUPARCO, an SLV that would place lightweight satellites into LEO. The proposed SLV would be derived from an already available missile launching system.[224] Another renowned nuclear scientist echoed the same pitch in August 2009 and added that the first indigenous satellite would be launched in 2011.[225] The indigenous SLV appears to be nowhere on the horizon. Achieving success with solid-fuelled military missiles has not conferred on Pakistan the technology required for launching satellites. Pakistan could well need to look at technology enhancements in the field of liquid-fuelled rockets so necessary for launching satellites and placing them in orbit.

The Satellite Arena

A shade more encouraging than the outcome of the indigenous SLV programme is Pakistan's progress, albeit slow, in the field of building satellites. The country has built three so far and launched them aboard foreign launchers. Development of its first satellite, the Badr 1, launched in 1990, was facilitated by a few engineers of SUPARCO who completed their master's degree in engineering from England's University of Surrey, the university that built and operated various small satellites.[226] It was in this university that the Pakistani engineers worked on UoSAT (University of Surrey SATellites that carried amateur radio transponders) projects and then, back home, along with some engineers at SUPARCO who were hams, commenced building a ham-satellite in 1986.[227] Christened Badr-1 (Urdu for new Moon), the design of the satellite was based on that of UoSAT-2[228] built by the University of Surrey in 1984. It took almost three decades after the establishment of SUPARCO for Pakistan's first satellite to be launched in 1990 aboard the Chinese Long March 2E rocket. It took another decade for the launch of the Badr-2 in December 2001 from Kazakhstan aboard the Russian Zenit 2 rocket. This was an imaging satellite in a Sun-synchronous 1,050 km altitude providing a resolution of approximately 250 m with a lifespan of two to three years.[229] The satellite was designed and built by the UK's Space Innovations, Ltd., with component integration done in Pakistan by SUPARCO.[230]

In the interim, in a move that would have forward-looking implications, Pakistan applied for five GEO slots from the International Telecommunications Union which were allotted to it in 1995. The progress on indigenous satellites or buying or leasing foreign satellites did not keep pace with the GEO slots reserved in advance, causing four of the five slots to expire. To prevent the loss of the last slot which would have expired in April 2003, Pakistan as an emergent measure, in 2002 leased the partially damaged, third-hand Palapa-C1 satellite. The Hughes-built Palapa-C1 satellite was launched for Indonesia in 1996 but developed electrical problems within three years, was bought by the insurance company that covered the project, then bought by Hughes Global Services in January 1999, leased to Turkey and placed in the allotted slot of 50 degree East in 2001, and then finally leased by Pakistan in 2002 for five years, as PAKSAT-1 in its slot of 38 degree East, at an initial cost of $ 4.5 million. The electrical problems came with the lease, which prevented power to the payload during the eclipse period of 88 days in a year. Buying the Palapa-C1 was then the cheapest face-saving option available to avoid losing the last of the five slots, prompting the Pakistani President to announce in 2003 plans for launching an indigenous Earth observation-cum-communications satellite within three years to replace the PAKSAT-1.[231] The aim was not only to preserve the GEO slot allotted to Pakistan, but also to speed up indigenising satellite technology so vital to its national prestige.

Pakistan's short-term ambition, however, remained that of observation and remote sensing. Not only would the dividends be of civilian applications including commercial but would, equally importantly, facilitate its military objectives of 'eyes in space'. An in-house study in 2007 recommended a constellation of optical and SAR satellites in the years ahead, as a part of high-resolution Remote Sensing Satellite System (RSSS) programme, commencing with an optical-payload remote sensing satellite planned for launch in 2011 that would provide 2.5 m panchromatic resolution from a 700 km Sun-synchronous orbit.[232] This satellite is yet to be launched but what it launched was PAKSAT-1R in August 2011. Manufactured by the China Great Wall Industry Corporation (CGWIC) based on the DongFangHong-4 satellite platform, the 5.2 ton telecommunications satellite launched aboard the Chinese Long March 3B/E into GEO has its electronics package manufactured in Pakistan with Chinese assistance. The satellite supports all the conventional and modern Fixed Satellite Services (FSS) applications via 18 transponders in the Ku- and 12 in the C-band. It is equipped with three receiver antennae and two transmission antennae that support 150-200 TV channels, Internet and data communications simultaneously to ground users who use a 45 centimetre antenna. With a design lifetime of 15 years, the PAKSAT 1R could well offer a broadened scope of applications in the future, as, in addition to high volume broadcast communications, the satellite can be used for data relay and as a regional mobile communication satellite.[233] Parked at the GEO slot of 38 degree East, PAKSAT 1R replaces PAKSAT 1 and provides coverage to over 75 countries in Europe, Africa, Middle East, South and Central Asia.[234] The launch follows almost immediately the approval of Pakistan's *Space Programme 2040* by the National Command Authority during its 19th meeting presided over by the country's Prime Minister in July 2011.[235] The plans entail launching of five satellites in GEO and six in LEO for remote sensing by the year 2040.[236]

Organisation and Infrastructure

The range of space related activities that are undertaken by Pakistan are controlled by SUPARCO, headquartered at Karachi, with some additional facilities at the University of Punjab at Lahore. SUPARCO is a state agency with a high degree of autonomy and implements the policies laid out by the Space Research Council, premiered by the country's Prime Minister. The early years of Pakistan's interest in space related applications saw the establishment in 1976 of a mobile receiving station in Rawalpindi to receive Earth resource data from the US Landsat system. A permanent facility was set up in Rawat, near Islamabad, in 1989, to receive data, in addition, from the French SPOT system[237] that provides 10 m resolution imagery and the US National Oceanic and Atmospheric Agency (NOAA)[238] that provides a resolution of 1.1 km. This facility is one of the most advanced of its kind in the Asia-Pacific region. The acquisition zone of this facility is about 2,500 km around Islamabad and its products are provided to more than 70 public and private enterprises in Pakistan. As the country's coordinator for remote sensing, SUPARCO trains users of this application in resources, urban planning and environment monitoring. In respect of training a qualified cadre of space professionals, SUPARCO has opened an Institute of Space Technology in Islamabad.[239] Design and development of satellites takes place under the auspices of the Satellite Research and Development Centre (SRDC) at Lahore while development of its sounding rockets occurs at the SUPARCO Plant at Karachi. With the new PAKSAT 1R in orbit with a lifespan of 15 years, Pakistan rightfully so, has aimed for the future by establishing in addition, two modern satellite ground control stations at Karachi and Lahore, one a back-up to the other.

At the international level, Pakistan has taken a lead in establishing an inter-state and non-political[240] Inter-Islamic Network on Space Sciences and Technology (ISNET) in 1986 to promote the advancement of space science and technology in the member countries of the Islamic world. It is presided over by the Chairmen of SUPARCO.

Pakistan's Motives and Space Drivers

Historically, Pakistan has accorded a higher priority to its nuclear programme and the development of its solid-fuel ballistic missiles. Among countries that possess nuclear technology, Pakistan has the weakest space capabilities after North Korea.[241] Though it had a space programme running for the past five decades, commencing with rather benevolent assistance from the US and Europe, further technological assistance from them trickled to almost a stop due to two reasons: firstly, the sanctions imposed on SUPARCO in the 1990s for its activities in missile proliferation and, secondly, A. Q. Khan's duplicity in stealing critical nuclear designs and knowhow from the Urenco enterprise in the Netherlands in pursuit of Pakistan's nuclear bomb becoming public knowledge.[242]

Pakistan's continuing its space ambitions despite these setbacks, arises from essentially two facets of the geopolitical placement of the country. First and foremost is the Indian equation. Seemingly overshadowed by the larger and militarily mightier arch-rival, India, Pakistan has typically vowed to match India in all of its military ventures.[243] This narrative remains equally true in the realm of space. The second motive lies in the position Pakistan sees itself in with regard to other Islamic states. Having stamped itself as the leader in the nuclear field in the Islamic world, Pakistan would naturally have political ambitions of achieving and maintaining this status in the field of space too, given the fact that Pakistan established ISNET way back in 1986.

To conclude the narrative of Pakistan's space ambitions, it would suffice to say that the same can only come about with a tremendous amount of technological advancements and the concomitant financial backing being accorded to it, which would, perforce, be at the cost of other more urgent nation-building domestic programmes. As Pakistan's space technology stands today, with non-affordable costs, limited indigenous expertise, sanctions on technology imports and a spoilt international reputation, Pakistan is most likely to remain a peripheral space power in the times to come.[244] In the Indian context, in times of crisis, the 'China factor' is likely to play a rather larger-than-supportive role for Pakistan. Even without China's overt participation, Pakistan could well piggyback on its ally's space wagon and receive for itself ready to use satellite-derived information, including tactical information in the space-support domain of its land wars.

Conclusion

Down the decades, countries that have exhibited space ambitions have focussed on three distinct though inter-related goals: scientific-technological progress, international prestige and national security.[245] Each spacefarer prioritised one over the other to suit its individualistic national and regional charter. Some have grander ambitions than the others and they achieved the status of being *space-capable* while others appeared content for the time being with being *space-enabled*, in that, these countries currently lack the technological resources of building indigenous satellites and launching them, and rely upon outright purchase or the collaborative approach with other space-capable countries for benefits to their populace and the economy.

The latecomers, the countries that followed the USSR and USA to join the space bandwagon, had certain advantages already accrued to them vis-a-vis the 'first wave' states. By the time the 'second wave' and then the 'third wave' states entered space, a certain amount of Soviet and American technology was already available. So was the knowledge on which areas of space technology offered the best dividends for a country's specific needs, negating wasting effort and resources on unproductive projects. Another benefit to latecomers is that disciplines in space sciences and engineering came to be included in specialised educational institutes for advanced training.[246] Such countries also have widespread access to space technologies available in the commercial market like GPS, weather data, remote sensing and information on space debris.[247] This empowers the customer-countries to build upon what is commercially available and enables them to prioritise and focus on indigenous capability building on facets suited to their national needs.

With regard to a country's individualistic motive for venturing into space, the USSR and USA set the trend with their military-driven space programmes. Other than these two, China is one which has a demonstrated military capable space programme. Origins of some of the countries' space programmes have had military roots, notably those of Israel, Iran, North Korea and Pakistan. In the case of the latter two, it was aimed to counter their larger and more capable neighbours. In the case of Israel, it is a singular issue of the country's survival. On the other hand, some countries like Japan and India initiated their nationalistic pride-driven space ventures purely for peaceful purposes and have only recently turned to military uses. Subsequent technological prowess emboldened China to pursue manned missions and some countries like China, Japan and India, to explore space beyond the Earth's immediate vicinity.

In terms of international cooperation in respect of the Asian space giants, that is, China, Japan and India, one issue is conspicuous by its absence. While all of them relied on various levels of foreign support from providers of space technology, all of them in almost equal measure maintained contacts and provided assistance to weaker regional partners for building political and economic influence. What remains missing is the linkage between and amongst the three.[248] The continent has not seen signs of any significant or substantive cooperation among the three in the realm of space technology. Is this a pointer to, and a harbinger of, an Asian space race in the years to come? Are the reasons purely of competition – in the domain of regional ascendency, political influence, or commercial interests? What is the likelihood of the space competition flowing out into the global arena? The coming years would need to see concerted efforts being made, at both the international and regional levels, to ensure that 'every country's right to space' without conflict of interest with any other space-user, within the mandate of international laws and agreements is facilitated.

Notes

1. France, on November 26, 1965, launched its first satellite, the Asterix, on its own rocket, the Diamant A, from Algeria, making it the third country in the world to do so after the USSR and the USA. The UK (1962), Canada (1962) and Italy (1964) too launched their very own first satellites but on American launch vehicles. It was not until October 1971 that the UK launched its satellite, the Prospero, on its own rocket, the Black Arrow.
2. Joseph A. Angelo, Jr, *Encyclopedia of Space and Astronomy* (New Delhi: Viva Books Private Limited), 2010, p. 220.
3. Ibid., p. 220.
4. http://www.esa.int/SPECIALS/About_ESA/SEMSN26LARE_0.html, accessed on April 2, 2012.
5. Keith Hayward, "Europe in Space*,"* *The RUSI Journal*, 155:6, 2010, pp. 36-42.
6. James Clay Moltz, *Asia's Space Race: National Motivations, Regional Rivalries And International Risks* (New York: Columbia University Press, 2012), pp. 1-2.
7. http://www.esa.int/SPECIALS/About_ESA/SEMFEPYV1SD_0.html, accessed on April 2, 2012.
8. Brian Harvey, *The Japanese and Indian Space Programmes: Two Roads into Space* (Chichester, UK: Praxis Publishing Ltd, 2000), p. 1.

9. Steven Berner, *Japan's Space Program: A Fork in the Road*? (Santa Monica, California: RAND, 2005), p. 3.
10. Berner, n. 9, p. 6.
11. Harvey, n. 8, p. 59.
12. Berner, n. 9, p. 14.
13. http://www.jaxa.jp/projects/rockets/h2a/index_e.html, accessed on April 4, 2012.
14. http://www.jaxa.jp/projects/sat/ikaros/index_e.html, accessed on April 9, 2012.
15. http://www.jaxa.jp/collabo/int/index_e.html, accessed on April 9, 2012.
16. http://www.jaxa.jp/press/2010/09/20100930_iac_e.html, accessed on April 9, 2012.
17. http://www.aprsaf.org/about/, accessed on April 9, 2012.
18. http://afe.easia.columbia.edu/special/japan_1950_usjapan.htm, accessed on April 10, 2012.
19. Berner, n. 9, p. 17.
20. Stephen Clark, "Japan to Build Fleet of Navigation Satellites," April 4, 2013, accessed at http://www. spaceflightnow.com/news/n1304/04qzss/#.UZtQy_i6YdU on May 21, 2013.
21. http://www.jaxa.jp/projects/sat/qzss/index_e.html, accessed on April 10, 2012.
22. Wg Cdr Kaza Lalitendra, *Militarisation of Space* (New Delhi: KW Publishers Pvt Ltd, 2010), p. 133.
23. Moltz, n. 6, pp. 9-10.
24. "Japanese MoD Unveils First Ever Space Budget," *Space News,* September 8, 2009, p.8.
25. Berner, n. 9, pp. 19-20.
26. Ajey Lele, *Asian Space Race: Rhetoric or Reality* (New Delhi: Springer, 2013), Ch. 17, p. 265.
27. Yasuhiro Morita, "A New Type of Launch Vehicle: A Rocket with Artificial Intelligence," accessed at http://www. jaxa.jp/article/interview/vol58/index_e.html on July 2, 2013.
28. http://www.jaxa.jp/projects/sat/sds4/index_e.html, accessed on April 11, 2012.
29. http://www.jaxa.jp/projects/sat/bepi/index_e.html, accessed on April 11, 2012.
30. http://www.jaxa.jp/projects/sat/hayabusa2/index_e.html, accessed on April 11, 2012.
31. Moltz, n. 6, p. 43.
32. Ibid., pp. 61-63.
33. Claire Noland in an article on Tsien's death on October 31, 2009, titled "Qian Xuesen Dies at 98; Rocket Scientist Helped Establish Jet Propulsion Laboratory" in *Los Angeles Times* of November 1, 2009, wrote about the event, "No one then knew that the father of the future U.S. space program was being quizzed by the father of the future Chinese space program".
34. Brian Harvey, *China's Space Program: From Conception to Manned Spaceflight*

(Chichester, UK: Praxis Publishing Ltd, 2004), pp. 17-20.

35. Harvey, n. 34, p. 22.
36. Roger Handberg and Zhen Li, *Chinese Space Policy* (London: Routledge, 2007), p. 57.
37. Harvey, n. 34, p. 27.
38. Ibid., p.37.
39. The DF 4 and the DF 5 comprise the mainstay of China's nuclear strike forces.
40. The text of the White Papers can be accessed at http://www.gov.cn/english/links/whitepapers.htm
41. Harvey, n. 34, pp. 32-33.
42. http://www.cnsa.gov.cn/n615709/n620681/n771967/69198.html, accessed on April 30, 2012.
43. Moltz, n. 6, pp. 88-89
44. http://www.cnsa.gov.cn/n615709/n620681/n771918/index.html, accessed on May 4, 2012.
45. Harvey, n. 34, pp. 153-157. Brazil had an active ongoing satellite programme, with its first satellite launch in 1993, but had not met with much success in its launcher programme, leading to this collaborative 70:30 ratio project between China and Brazil.
46. *Annual Report on the Military Power of PRC-2004,* at http://www.thespacereview.com/archive/1155.pdf, accessed on May 17, 2012.
47. Lalitendra, n. 22, pp. 107-108.
48. Ministry of Science and Technology of the PRC, *Newsletter*, No 441, June 10, 2006, accessed on May 8, 2012 at http://www.most.gov.cn/eng/newsletters/2006/200606/t20060613_33933.htm.
49. http://www.dragoninspace.com/rocketry/default.aspx, accessed on May 8, 2012.
50. "Annual Report on the Military Power of PRC-2000" at http://www.thespacereview.com/archive/1155.pdf, accessed on May 17, 2012.
51. http://www.bbc.co.uk/news/technology-16337648, accessed on May 2, 2012.
52. http://www.nasaspaceflight.com/2012/04/china-launch-compass-duo-long-march-3b/, accessed on May 2, 2012.
53. Harvey, n. 34, p. 248.
54. Stephen Clark, "China Says it has Spent Six Billion Dollars on Human Spaceflight," June 26, 2012, accessed at http://www.spaceflightnow.com/china/shenzhou9/120626cost/ on June 27, 2012.
55. Shenzou spacecraft looks similar to the Soviet Soyuz spacecraft, though it is longer and has a larger habitable volume.
56. Some other technologies imported from Russia included docking module, rendezvous systems, space station technology, life support systems and space suits. This makes

the Shenzou spacecraft, like the Soyuz, compatible for docking with the International Space Station.

57. Article on China's rockets at http://www.dragoninspace.com/rocketry/default.aspx, accessed on May 9, 2012.
58. http://www.space.com/14048-china-satellite-launch-breaks-rocket-record.html, accessed on May 8, 2012.
59. "Space Launch Report," accessed at http://www.spacelaunchreport.com/log2012.html on July 9, 2013.
60. Remarks of Bao Shixiu in "Deterrence Revisited: Outer Space," *China Security*, vol. 3, no. 1, Winter 2007.
61. Monika Chansoria, "'Informationising' Warfare: China Unleashes the Cyber and Space Domain," *Manekshaw Paper,* no. 20, 2010, p. 28.
62. J. C. Liou and N. L. Johnson, "Physical Properties of the Large Fengyun-1 Breakup Fragments," *Orbital Debris Quarterly 12,* no. 2, April 2008, p. 4.
63. "This (the Chinese ASAT test) was absolutely preventable. The Chinese have been proposing a treaty to ban weapons in space for years. We have refused in order to pursue this fantasy of space-based antimissile weapons..." commented Joseph Cirincione of the Centre for American Progress, a research group, quoted in an article "U.S. Knew of China's Missile Test, but Kept Silent" in *New York Times* of April 23, 2007, accessed on May 10, 2012 at http://www.nytimes.com/2007/ 04/23/ washington/23satellite.html.
64. Cited in Marc Kaufman and Dafna Linzer, "China Criticized for Anti-Satellite Missile Test," *Washington Post*, January 19, 2007. Michael Krepon is co-founder of the Stimson, Washington, a non-profit institution devoted to enhancing international peace and security, and the Director of its South Asia and Space Security Programmes.
65. Quoted by Kevin Pollpeter in Tania Branigan, "China Launches Manned Spacecraft," *The Guardian,* September 26, 2008. Kevin Pollpeter is China Programme Manager at the Defense Group Inc.'s Centre for Intelligence Research and Analysis, Washington, DC. Previously, he was a researcher at the RAND Corporation. Mr. Pollpeter has published widely on China's national security issues and focusses on the Chinese space programme.
66. Michael Sheehan, *The International Politics of Space* (London: Routledge, 2007), p. 161.
67. Moltz, n. 6, pp. 91-92.
68. http://www.dragoninspace.com/Earth-observation/ziyuan3.aspx, accessed on May 14, 2012.
69. China's White Paper of 2011 on space, lists the role of the Tianhui as 'space mapping'.
70. http://www.dragoninspace.com/Earth-observation/ziyuan3.aspx, accessed on May 14, 2012.

71. "China no Longer Reliant on Satellite Image Imports," accessed at http://www.spacedaily.com/reports /China_no_longer_reliant_on_satellite_image_imports_999.html on April 25, 2013.
72. http://www.dragoninspace.com/communications/fenghuo.aspx, accessed on May 14, 2012.
73. Wg. Cdr. Kaza Lalitendra, "Dragon in Space: Implications for India," *Air Power*, vol. 3, no. 3, Monsoon 2008, p. 95.
74. Mark A. Stokes, *China's Strategic Modernization: Implications for the United States* (Carlisle, PA: US Army War College-Strategic Studies Institute, 1999), p. 118.
75. Ashley J. Tellis, "China's Military Space Strategy," *Survival*, vol. 49, issue. 3, 2007, p. 53 accessed on May 14, 2012 at http://www.carnegieendowment.org/files/tellis_china_space1.pdf
76. Ibid., p. 54.
77. Transcript of the testimony of Gen James E. Cartwright, Commander, US Strategic Command, before the Strategic Forces Subcommittee of the Senate Armed Services Committee, Russell Senate Office Building, Washington DC, March 28, 2007, excerpted from Ibid., p. 54.
78. Leonard David, "Shenzhou-7 Deployed Picosat That Came Within 25 Kilometers of Space Station," *Space News,* October 20, 2008, p. 10.
79. Richard J. Adams and Martin E. France, "The Chinese Threat to U.S. Space Superiority," *High Frontier*, vol. 1, no. 3, Winter 2005, pp. 18-23.
80. *Annual Report on the Military Power of PRC-2003,* at http://www.thespacereview.com/archive/1155.pdf, accessed on May 17, 2012.
81. Vishal Nigam, "Transformational Trends in China's Aviation Industry," *Asian Defence Review 2011,* p. 53.
82. "China Eyes 15 pct Market Share in Commercial Space Launch," March 2, 2013 accessed at http://news. xinhuanet.com /english/china/2013-03/02/c_132203004.htm on July 15, 2013.
83. Lele, n. 26, Ch. 17, p. 265.
84. "Long March 5 to Take Off in 2014" at http://www.chinadaily.com.cn/china/2012-03/03/content_14748197.htm, accessed on May 18, 2012.
85. http://www.spacedaily.com/reports/Long_March_5_Will_Have_World_Second_Largest_Carrying_Capacity_999.html, accessed on May 18, 2012.
86. http://www.spacedaily.com/reports/China_to_launch_100_satellites_during_2011_15_999.html, accessed on May 18, 2012.
87. "Space Launch Report," accessed at http://www.spacelaunchreport.com/log2012.html on July 9, 2013.
88. Leonard David, "China Details Ambitious Space Station Goals", accessed at http://www.space.com/11048-china-space-station-plans-details.html on May 21, 2012.

89. Julie Wakefield, "Moon's Helium 3 Could Power Earth," accessed at http://www.helium3.ru/articles/28.pdf on May 21, 2012.
90. "Transforming Adversity to Advantage", accessed at http://www.mfa.gov.il/NR/rdonlyres/DB9BB989-8888-41CB-926C-AA9EE46D8397/0/ch78.pdf on May 22. 2012.
91. Brian Harvey, Henk H. F. Smid and Theo Pirard, *Emerging Space Powers: The New Space Programmes of Asia, the Middle East, and South America* (Chichester, UK: Praxis Publishing Ltd., 2010), pp. 385-387.
92. "Israel: How Far Can its Missiles Fly?" *The Risk Report*, vol. 1, no. 5, p. 1.
93. "Dassault Lifts the Lid on the Jericho Missile Story," *Air and Cosmos/Aviation International,* December 6, 1996, p. 36.
94. Henry A. Kissinger's memorandum dated July 16,1969, to the US President, accessed on May 22, 2012, at http://nixon.archives.gov/virtuallibrary/documents/mr/071969_israel.pdf
95. "Shavit (Israel), Space Launch Vehicles- Orbital," excerpted from http://articles.janes.com/articles/ Janes-Space-Systems-and-Industry/Shavit-Israel.html on May 23, 2012.
96. Ames Dunnigan, "Jericho 3 Definitely in Service," at http://www.strategypage.com/dls/articles/ 20082522184.asp, accessed on May 24, 2012.
97. Daphne Burleson, *Space Programmes Outside the United States* (North Carolina, USA: McFarland & Company Inc. Publishers, 2005), pp. 153-157.
98. Harvey, et al., n. 91, p. 387.
99. Ibid., n. 91, p. 398.
100. Eric Silver, "Questions and Answers with ISA Chairman, Isaac Ben Israel," accessed on May 25, 2012 at http://www.mfa.gov.il/NR/rdonlyres/A7C494F2-62C2-44BC-8FA1-148D776A67DA/0/ch76.pdf
101. Professor M.G. (Retd) M.K. Isaac Ben-Israel and Dr Zvi Kaplan," New Frontiers-Out of this World: Israel's Space Programme," accessed on May 25, 2012 at http://www.mfa.gov.il/NR/rdonlyres/A7C494F2-62C2-44BC-8FA1-148D776A67DA/0/ch76.pdf
102. http://www.most.gov.il/English/Units/Israel+Space+Agency/ISA+Activity/Satellites.htm accessed on May 25, 2012.
103. http://www.spacelaunchreport.com/slr2010q2.html, accessed on May 25, 2012.
104. "High Tech Sat for Alleged Israeli Raid," September 17, 2007, accessed at http://www.warisboring.com/2007/09/17/hiigh-tech-sat-for-alleged-israeli-raid/ on May 30, 2012.
105. Harvey, et al., n. 91, p. 432.
106. Ibid., n. 91, pp. 401-404.
107. "India Launches Israeli Satellite," January 21, 2008, accessed at http://news.bbc.

co.uk/2/hi/south_asia /7199736. stm on May 31, 2012.

108. Tzvi Ben Gedalyahu, "Kazakhstan Launches Israel's Amos 5 Communications Satellite" accessed on May 28, 2012 at http://www.israelnationalnews.com/News/News.aspx/150605.
109. David A. Fulghum, "Micro and Nano Sats," *Aviation Week and Space Technology,* June 16, 2003, p. 140.
110. Barbara Opall Rome, "Israel Studies Airborne Launch Scheme for Shavit Rocket," July 22, 2011, accessed at http://www.spacenews.com/launch/110722-israel-airborne-launch-shavit.html on May 30, 2012.
111. Harvey, et al., n. 91, p. 433.
112. http://www.most.gov.il/English/Units/Israel+Space+Agency/ISA+International+Relations.htm, accessed on May 29, 2012.
113. Eric Silver, "Questions and Answers with ISA Chairman, Isaac Ben Israel," accessed on May 25, 2012 at http://www.mfa.gov.il/NR/rdonlyres/A7C494F2-62C2-44BC-8FA1-148D776A67DA/0/ch76.pdf
114. Lele, n. 26, Ch. 17, p. 197.
115. Arie Egozi, "IAI Planning Construction of Six New Satellites," dated January 30, 2012, accessed at http://www. israeldefense.com/?CategoryID=474&ArticleID=866 on May 30, 2012.
116. Amir Rapaport and Nir Dvori, "Israel will Concentrate on Sophisticated Communications and Microsatellites," dated June 20, 2011, accessed at http://www. israeldefense.com/?CategoryID=472&ArticleID=466 on May 30, 2012.
117. Barbara Opall Rome, "Israel Studies Airborne Launch Scheme for Shavit Rocket," July 22, 2011, accessed at http://www.spacenews.com/launch/110722-israel-airborne-launch-shavit.html on May 30, 2012.
118. Stephen Clarke, "Iran Launches Omid Satellite Into Orbit," accessed at http://www. space.com/5432-iran-launches-omid-satellite-orbit.html on June 4, 2012.
119. Intelsat was founded as a public-private consortium in 1964 by the telecommunication agencies of 18 nations, including the USA which proposed the organisation, to provide satellite communications services. The transmitting and receiving apparatus in each country was owned by the Intelsat member, the telecommunication agency from that country. The consortium contracted NASA to launch its satellites. The first of these satellites was the Early Bird, later renamed Intelsat I, which was placed in a stationary orbit over the Atlantic Ocean at the Equator in 1965. The Early Bird was the first operational commercial satellite, providing regular telecommunications and broadcasting services between North America and Europe and it was Intelsat III F-3 satellite that beamed the Moon landing of 1969 live to 600 million TV viewers across the globe.
120. Harvey, et al., n. 91, pp. 255-257.

121. http://www.spacedaily.com/news/iran-05d.html accessed on June 5, 2012.
122. Ameer Alam, "The Emerging Iranian Space Programme" accessed at http://www.pakalumni.com/profiles/blogs/ 1119293:BlogPost:64819 on June 5, 2012.
123. Sharon Weinberger, "Russians Say 'Nyet' To Another Iranian Satellite," June 25, 2009, accessed at http://www. wired.com/dangerroom/2009/06/russians-say-nyet-to-another-iranian-satellite/ on June 5, 2012.
124. Harvey, et al., n. 91, pp. 258-262.
125. Yiftah Shapir, "Iran's Effort to Conquer Space, Strategic Assessment," The Institute for National Security Studies, vol 8, no. 3, November 2005, accessed at http://www.inss.org.il/publications.php?cat=25&incat=&read=160 on June 6, 2012.
126. Parviz Tarikhi, "*Iran's Space Ambitions,*" May 13, 2010, accessed at http://www.spatialsource.com.au/article /Irans-Space-Ambitions/ZETSPMAZSF.html?page=4 on June 6, 2012.
127. http://www.iranwatch.org/suspect/records/aerospace-research-institute.html accessed on June 6, 2012.
128. There are over two dozen Iranian universities that provide space related graduate and postgraduate degree education.
129. http://www.ieimil.ir/content/about-us, accessed on June 6, 2012.
130. http://www.iranwatch.org/suspect/records/iran-electronics-industries-(iei).html accessed on June 6, 2012.
131. IEI and some of its subsidiaries/front companies had sanctions placed against them by the EU and USA in 2008 for proliferation of weapons of mass destruction and nuclear material.
132. Harvey, et al., n. 91, pp. 300-301.
133. "Iranian Official: Italy Deviates from Handing over Misbah Satellite," April 5, 201, accessed at http://en.trend.az/regions/iran/1856293. html#popupInfo on June 6, 2012.
134. Yiftah Shapir, "Iran's Effort to Conquer Space, Strategic Assessment," The Institute for National Security Studies, vol 8, No. 3, November 2005, accessed at http://www.inss.org.il/publications.php?cat=25&incat=&read=160 on June 6, 2012.
135. Harvey, et al., n. 91, pp. 298-300.
136 President Mahmoud Ahmadinejad's post-launch announcement, accessed at http://www.isa.ir/components1 .php?rQV===AfABkO0hXZUh2YyFWZz9lZ8BUMApD ZJRnblJXYw
9lZ8BEM5EDQ6QWStVGdp9lZ8BUM4ATMApDZJ52bpR3Yh9lZ on June 7, 2012.
137. "Iran's Latest Satellite Named Rasad was Launched into Earth Orbit," accessed at http://www.isa.ir/ components1.php?rQV===AfABkO0hXZUh2YyFWZz9lZ8BU MApDZJRnblJX

Yw9lZ8BkN5QDQ6QWStVGdp9lZ8BUM4ATMApDZJ52bpR3Yh9lZ on June 7, 2012.

138. Julian Borger, "Iran Launches Second Satellite into Orbit, Claims State TV," *The Guardian*, June 16, 2011, accessed at http://www.guardian.co.uk/world/2011/jun/16/iran-satellite-launch-orbit on June 7, 2012.

139. http://www.isa.ir/components1.php?rQV===AfABkO0hXZUh2YyFWZz9lZ8BUMApDZJRnblJXYw9lZ8B0N3MDQ6QWStVGdp9lZ8BUM4ATMApDZJ52bpR3Yh9lZ, accessed on June 7, 2012.

140. "US Calls Iran's *Kavoshgar* Rocket Launch 'Provocative Act'," February 3, 2010, accessed at http://en.rian.ru/ world/20100203/157765694.html on June 7, 2012.

141. Harvey, et al., n. 91, pp. 306-307.

142. "Another Honor for the Islamic Republic of Iran," accessed at the ISA site at http://www.isa.ir/ components1.php?rQV==wHQyAkOklUZnFWdn5WYMJXZ0VWbhJXYw9lZ8B0M1UDQ6QWStVGdp9lZ8BUM4ATMApDZJ52bpR3Yh9lZ on June 7, 2012.

143. "Iran Successfully Launches New Satellite into Orbit," February 3, 2012, accessed at http://www.presstv.ir/detail/ 224625.html on June 7, 2012.

144. Harvey, et al., n. 91, pp. 285-294.

145. Charles P. Vick, "Iran Space Launch Vehicles," May 25, 2010, accessed at http://www.globalsecurity.org/ space/world/iran/launch.htm on June 7, 2012.

146. http://www.astronautix.com/lvs/safir2.htm, accessed on June 7, 2012.

147. Charles P. Vick, "Safir 1, 2 - Iran First Satellite Booster," August 6, 2010, accessed at http://www.globalsecurity. org/ space/world/iran/safir-1-2.htm# on June 7, 2012.

148. Charles P. Vick, "Safir Block II- Second Generation Class Space Booster & Future Booster Developments," May 25, 2010, accessed at http://www.globalsecurity.org/space/world/iran/safir-block2.htm on June 8, 2012.

149. Charles P. Vick, "Simorg-3 IRILV Booster Family," September 1, 2010, accessed at http://www.globalsecurity.org/ space/world/iran/simorgh-3-series.htm on June 8, 2012.

150. Ibid.

151. "Update: Iran Expands Missile Arsenal," March 8, 2010, accessed at https://www.thetrumpet.com/ ?q=7028.5562.0.0 on June 7, 2012.

152. "Iran Unveiled New Generation of Satellites," February 3, 2010, accessed at http://www.isa.ir/components1. php?rQV===AfABkO0hXZUh2YyFWZz9lZ8BUMApDZ JRnblJXYw9lZ8BkN3MDQ6QWStVGdp9lZ8BUM4ATMApDZJ52bpR3Yh9lZ on June 8, 2012.

153. n.152.

Gdp9lZ8BUM4ATMApDZJ52bpR3Yh9lZ on June 8, 2012.

154. "Iran to Launch Toloo Satellite Next Year," accessed at http://www.payvand.com/

news/11/nov/1025.html on June 8, 2012.

155. http://www.presstv.ir/detail/165867.html, accessed on June 8, 2012.
156. Harvey, et al., n. 91, p. 277.
157. http://www.satlaunch.net/2011/09/iran-to-launch-qaem-satellite-in-2017.html, accessed on June 8, 2012.
158. "Report: Iran Unveils Self-Made Satellite Navigation System," May 31,2012, accessed at http://www.haaretz. com/news/middle-east/report-iran-unveils-self-made-satellite-navigation-system-1.433673 on June 8, 2012.
159. "Iran to Implement 10 Satellite Projects with APSCO Members," October 20, 2010, accessed at http://www. payvand .com/news/10/oct/1160.html, on June 8, 2012.
160. Harvey, et al., n. 91, pp. 305-306.
161. "Iran Plans Manned Space Mission in the Next 10 Years," *Reuters*, August 21, 2008, accessed at http://www.reuters. com/article/2008/08/21/us-iran-space-idUSBLA13323720080821 on June 11, 2012.
162. Harvey, et al., n. 91, p. 308.
163. Accessed at http://www.bbc.co.uk/news/world-asia-17698438 on June 18, 2012.
164. Harvey, et al., n. 91, p. 439. Besides the US' impressive space surveillance capability based on a worldwide network of radars and optical sensors to detect a missile launch within seconds and then track the spacecraft up to its orbit altitudes, radio-monitoring of 27 MHz frequency also detected no transmissions from the 'ghost-satellite', notwithstanding the DPRK's claims of an onboard transmitter operating at the citizen-band frequency of 27 MHz.
165. Brian Weeden, "The Truth about the North Korean Space Launch," December 19, 2012, accessed at http://www.wired.co.uk/news/archive/2012-12/19/north-korean-satellite on July 8, 2013.
166. Harvey, et al., n. 91, pp. 439-441.
167. Burleson, n. 97, pp. 206-209.
168. "North Korea Missile Milestones-1969-2011," accessed at http://www. wisconsinproject.org/countries /nkorea/north-korea-missile-milestones.html on June 20, 2012.
169. "North Korea's Missile Programme," May 29, 2009, accessed at http://news.bbc. co.uk/2/hi/asia-pacific/ 2564241.stm on June 20, 2012.
170. "North Korea's Missile Programme," May 29, 2009, accessed at http://news.bbc. co.uk/2/hi/asia-pacific/ 2564241.stm on June 20, 2012.
171. Charles P. Vick, "Taep'o-dong-3," May 25, 2012, accessed at http://www. globalsecurity.org/wmd/ world/ dprk/td-3.htm# on June 21, 2012.
172. Lewis Franklin and Nick Hansen, "North Korea's Space Programme," *Jane's Intelligence Review,* September 2009, p. 12.
173. Terence Roehrig, "North Korea in Crises: Regime, Identity, and Strategy," in Shale

Horowitz, Uk Heo and Alexander C. Tan, eds., *Identity and Change in East Asian Conflicts: The Cases of China, Taiwan, and the Koreas* (New York: Palgrave, 2007), p. 142.

174. Moltz, n. 6, pp. 171-173.
175. Ibid., p. 141.
176. Wyn Bowen, *The Politics of Ballistic Missile Proliferation* (London: MacMillan Press, 2000), p. 34.
177. http://www.astronautix.com/country/korsouth.htm, accessed on June 14, 2012.
178. http://www.fas.org/irp/threat/missile/skorea.htm, accessed on June 13, 2012.
179. Moltz, n. 6, p. 145.
180. Report on Korea's National Space Programme (2000-2015), *Publication of National Science and Technology Council,* Republic of Korea.
181. Harvey, et al, n. 91, p. 512.
182. Moltz, n. 6, pp. 145-146.
183. Somi Seong and Steven W. Popper, *Strategic Choices in Science and Technology: Korea in the Era of a Rising China* (Santa Monica, California: RAND, 2005), pp. 103-104.
184. Harvey, et al., n. 91, p. 492.
185. "Korea to Launch First Military Satellite in 2006," *Korea Herald,* November 18, 2002.
186. Not surprisingly, KoreaSat-4 appears to have been missed in this series. The number 'four' in the Sino-Korean language is 'sa' which is a homonym for the Chinese character for death and usually avoided, akin to the number '13' in the Western world.
187. www.thalesgroup.com/Case_Studies/Documents/koreasatv_en_pdf/, accessed on June 15, 2012.
188. http://space.skyrocket.de/doc_sdat/koreasat-6.htm, accessed on June 15, 2012.
189. "Multipurpose Satellite Arirang 2 set for Launch," *Korea Herald,* July 28, 2006.
190. http://astronomy.activeboard.com/t7638884/arirang-2-satellite/, accessed on June 15, 2012.
191. Jessica Seoyoung Choi, "Arirang 3 in Space," May 19, 2012, accessed at http://m.korea.net/english /NewsFocus/Sci-Tech/view?articleId=100434 on Jun 15, 2012.
192. Harvey, et al., n. 91, p. 502.
193. Moltz, n. 6, pp. 136-148.
194. Lee Joo-jin, KARI President, 2009, quoted in Moltz, n. 6, p. 148.
195. Moltz, n. 6, p. 156.
196. http://www.astronautix.com/lvs/kslvii.htm accessed on June 15, 2012.
197. Won-hwa Park, "Report: Recent Space Developments in South Korea," *Space Policy,* vol. 26, no. 2, May 2010, p. 118.

198. Moltz, n. 6, p. 155.
199. Daniel A. Pinkston, "North and South Korean Space Development: Prospects for Cooperation and Conflict," *Astropolitics,* vol. 4, no. 2, Summer 2006, p. 221.
200. Moltz, n. 6, pp. 165-166.
201. http://www.satbeams.com/satellites?norad=35812, accessed on June 26, 2012.
202. Burleson, n. 97, p. 148.
203. http://www.globalsecurity.org/space/world/indonesia/eo.htm, accessed on June 26, 2012.
204. http://www.astronautix.com/lvs/rx250lpn.htm, accessed on June 27, 2012.
205. http://www.allmilitaryweapons.com/2009/09/indonesia-succesfully-launches-rocket.html, accessed on June 27, 2012.
206. http://www.globalsecurity.org/space/world/indonesia/rps.htm, accessed at June 27, 2012.
207. "RX 420 (Satellite Launch Vehicle)," accessed at http://madeinindonesiapedia.wordpress.com/2011/10/06/ rx420/, on June 27, 2012.
208. Moltz, n. 6, pp. 168-169.
209. http://www.astronautix.com/project/measat.htm, accessed on June 27, 2012.
210. http://www.astronautix.com/country/malaysia.htm, accessed on June 27, 2012.
211. https://www.cia.gov/library/publications/the-world-factbook/rankorder/2001rank.html, accessed on June 27, 2012.
212. Klaus Schmidt, "SpaceX Falcon 1 RazakSat Launch Successful," July 14, 2009, accessed at http://spacefellow ship.com/news/art10288/spacex-falcon-1-razaksat-launch-successful.html on June 27, 2012.
213. Moltz, n. 6, p. 168.
214. http://www.defenseindustrydaily.com/malaysia-receives-first-2-su30mkms-03336/ accessed on June 28, 2012.
215. http://www.geospatialworld.net/index.php?option=com_content&view=article&id=23030%3Amalaysia-emb arks -on-razaksat-2-programme&catid=46%3Aproduct-remote-sensing-satellite&Itemid=1, accessed on June 28, 2012.
216. http://www.suparco.gov.pk/pages/functions.asp, accessed on July 2, 2012.
217. "Pakistan Derives its First "Hatf" Missiles from Foreign Space Rockets," *The Risk Report*, vol. 1, no. 8, October 1995, p. 4, accessed at http://www.wisconsinproject.org/countries/pakistan/hatf.html on July 3, 2012.
218. Ibid.
219. The US sanctions were placed off and on, and finally lifted in 2010, to facilitate Pakistan's cooperation in USA's global fight against terror.
220. Sqn Ldr K.K. Nair, *Space: The Frontiers of Modern Defence* (New Delhi: Knowledge World, 2006), p. 136.
221. "China's Missile Sales to Islamabad Worry Washington," *The Risk Report,* vol. 1, no.

4, May 1995, accessed at http://www.wisconsinproject.org/countries/china/china-missile-sales.html on July 3, 2012.

222. Sqn Ldr Ajey Lele, "Pakistan's Space Capabilities," *Air Power,* vol. 2, no. 1, Spring 2005, p. 131.

223. "Pakistan Missile Milestones-1961-2005," *The Risk Report,* vol. 12, no. 1, February 2006, accessed at http://www.wisconsinproject.org/countries/pakistan/miss-miles2005.htm on July 3, 2012.

224. Nair, n. 220, p. 137.

225. "Pakistan's Shaheen III? Space Booster Development," accessed at http://www.globalsecurity.org/ space/world/pakistan/pakistan-shaheen-iii-space-booster.htm on July 3, 2012.

226. It is the University of Surrey that formed an offshoot company, the Surrey Satellite Technology Ltd. (now owned by EADS Astrium) that builds and operates small satellites, including South Korea's first satellite, the KITsat-1 in 1992 and China's first microsat, the Tsinghua-1 in June 2000.

227. Nair, n. 220, p. 134.

228. http://centaur.sstl.co.uk/SSHP/micro/micro90.html, accessed on July 4, 2012.

229. Lele, n. 222, pp. 135-145.

230. Nair, n. 220, p. 139.

231. Ibid., pp. 134-142.

232. http://www.suparco.gov.pk/pages/rsss.asp, accessed on July 5, 2012.

233. Rui C. Barbosa, "China Debuts Partnership with Pakistan - Long March Launches Paksat-1R," accessed at http://www.nasaspaceflight.com/2011/08/china-debuts-partnership-pakistan-long-march-launches-paksat-1r/ on July 5, 2012.

234. http://www.paksat.com.pk/, accessed on July 5, 2012.

235. http://www.thenews.com.pk/NewsDetail.aspx?ID=18716&title=NCA-okays-Nuclear-Power-Prog-2050, accessed on July 5, 2012.

236. Dr. Riaz Suddle, Member (Space Technology Wing), SUPARCO , "Pakistan's Space Programme- Opportunities for R&D," December 19, 2009, http://webcache.googleusercontent.com/search?q=cache:kyUXbO6sU8YJ:www.uet.edu.pk/export/sites/ UETWebPortal/newsannouncement/newssection/Workshop_WC_Dec_19_2009/Dr_Riaz_Suddle.ppt+M+Riaz+Suddle+electrical+engineering&cd=5&hl=en&ct=clnk&gl=us&source=www.google.com

237. French SPOT satellites are high resolution Earth observation satellites providing optical and multi-spectral imagery commercially to users across the globe.

238. NOAA is a scientific agency within the US Department of Commerce dealing with the conditions of oceans, weather and the atmosphere.

239. "SUPARCO," accessed at http://www.fas.org/spp/guide/pakistan/agency/index.html on July 5, 2012.

240. "About ISNET," accessed at http://www.isnet.org.pk/ on July 5, 2012.
241. Moltz, n. 6, p. 173.
242. For further reading on A. Q. Khan's activities, see David Albright, *Peddling Peril: How the Secret Nuclear Trade Arms America's Enemies* (New York: Free Press, 2010).
243. Moltz, n. 6, p. 176.
244. Lele, n. 222, p. 144.
245. Moltz, n. 6, p. 25.
246. Ibid., p. 24.
247. Ibid., p. 159.
248. Ibid., p. 33.

4

Space: The Indian Connect

As soon as we get over our present troubles, there will be a flowering of science and other activities in India which will probably astonish the world.

— Pandit Jawahar Lal Nehru
The first Prime Minister of Independent India,
in his address to the Indian Science Congress in 1947

India achieved its independence on August 15, 1947, after two centuries of British colonial rule and under the stewardship of its charismatic Prime Minister, Pandit Jawahar Lal Nehru, commenced its first steps towards nation building. The task of building a 'new India' right after the holocaust of partition was indeed a herculean one, given the fact that it involved binding together almost 400 million poverty-stricken Indians spread across a large and diverse landmass that makes it the seventh largest country in the world. Equally daunting was the task of restoring India's place in the economic world order, where India's share of the world income was a whopping 22.6 percent in 1700 before the colonial rule but a mere 3.8 percent on independence.[1]

History bears testimony to the route taken by poor countries to become richer, by relying on building institutions and banking on technology. The words spoken in the year of India's independence by Pandit Nehru to the Indian Science Congress duly recognised the role that science had to play and were prophetic enough to chart a course and set the pace for national development. Pandit Nehru, the architect of modern democratic and secular India, committed the country to freedom, socialism and social justice, backed by a solid institutional base aimed towards development. With the framework of a functional democracy having been established, the economy coasted along at the leisurely 'Hindu rate of growth' of 2.5-3 percent, but failed to attract any significant foreign trade and investment.

The early 1990s witnessed a major game changing shift in the national economic policy. The then Prime Minister P. V. Narasimha Rao along with his Finance Minister (and current Prime Minister) Man Mohan Singh opened the doors to economic liberalisation. Pandit Nehru's reliance on the public sector gave way to breaking the public sector monopoly[2] by privatising a number of state corporations. Controls on import, private enterprise and foreign investment were eased, custom duties lowered and the stifling licence *raj* significantly brought under check. More than 80 percent of Indian industry was delicensed, the number of industries reserved for the public sector was reduced from seventeen in 1990 to six, and plans were put in place for the divestment and privatisation of public sector undertakings.[3] Growth touched 7.5 percent in the mid-1990s, inflation came down by 7 percent and the ground was set to usher in an economic boom that preceded the Information Technology (IT) revolution and increase in foreign investment, as these were the years that coincided with the rise of Business Process Outsourcing (BPO) worldwide.

India rose from strength to strength and stands today as one the fastest growing and leading economies of the world and with a GDP of US$ 1.85 trillion in 2011 as per the World Bank,[4] has acquired a formidable status of economic strength that comes with concomitant geopolitical clout. Amidst its phenomenal growth narrative, India, a nation of paradoxes, has displayed an almost equal measure of stark contrasts. The economy stands as the ninth largest GDP in the world, a position that slips to the 129th when viewed against the per capita index. While India's seven poorest states remain home to more than half of the nation's poor, its higher-income states have successfully reduced poverty to levels comparable with some of the richer Latin American countries.[5] The success story in white collar sectors like IT, amongst others, has been a key driving force fuelling the economic growth and a hallmark of the country's achievement at the global level, exporting software to over 90 countries, earning for itself a brand identity as a knowledge-based economy. Backed by one of the largest pools of scientists and engineers in the world, the contrast is conspicuous when the country also has the highest high school drop-out rates in the world. Figures on poverty in 2010 convey almost 300 million subsisting on less than US$ 1.25 a day[6] while in the same year, 16 million tons of Indian-grown grain was lost to spoilage, enough to feed an estimated 118 million people for a year.[7] Its GDP is just under 3 percent of the world GDP, but India has nearly 7 percent of the world's 1,000 or so billionaires, whose combined wealth is more than 20 percent of the nation's GDP.[8] Sprawling mansions of the super-rich coexist with the squalor and poverty of impoverished slums in cities all over the country and further bring to the fore the visible diversity and mixed bag of fortunes that is contemporary India.

The 'Indian diversity' plays an equally significant part in the nation's geopolitics. Since its independence, it has anchored its foreign policy on peaceful coexistence and harboured friendly relations with its neighbours. Almost 45 percent of the over 15,000 km land border is contiguous with Pakistan and China, and territorial and border disputes have led to three full-fledged and one local war with Pakistan and one with China. This has necessitated maintaining a strong military, which, with a strength of 1.3 million, is the third largest in the world. Faced with the twin challenges of a militarily stronger nation to the north, which is also an emerging global economic and military superpower with expansionist tendencies, and an arch rival to the west which has also taken to subcontinental, and, of late, global terrorism as a political tool, India can ill afford any complacency in its security paradigm. It would be stating the obvious that India would need to exploit every dimension and instrument in its armoury of geopolitics, international relations, strategic considerations, economic influence, maritime and military capabilities, including the realm of space, to safeguard its legitimate security concerns. There are some who argue that India could have avoided the myopic Panipat syndrome, or lack of strategic vision, at least vis-a-vis Pakistan when India failed to capitalise on the gains achieved in its victory in the 1971 War. On the other hand, there are very few who acknowledge the fact that despite the bad blood across 45 percent of the nation's land borders, India has displayed tremendous acuity and maturity by not upping the ante, and in its ventures in space, the new dimension, took the path for peaceful purposes for the benefit of its people – a space programme with no content that offered any military utility.

India's first indigenous satellite, the Aryabhatta, entered orbit in April 1975, launched aboard the Soviet Cosmos 3M rocket. It took another five years for the nation to achieve its first indigenous space launch. The four-stage solid-fuel Satellite Launch Vehicle-3 (SLV-3) blasted off from the Indian launch site Sriharikota near its east coast in July 1980, successfully placing the 35 kg Rohini-1 satellite into a LEO, enabling the nation to become the seventh in the world to join the select group of 'second wave' of spacefaring countries – countries that achieved the capability of indigenous space launchers that could launch and place satellites in orbit. This was at a time when India had already acquired nuclear capability in 1974 but the space programme was conceived and executed with a purely peaceful orientation. This chapter chronicles the motives behind the Indian space programme and traces its evolution to the contemporary times.

Science in India Through the Ages

> *We owe a lot to the Indians, who taught us how to count, without which no worthwhile scientific discovery could have been made.*
>
> — Albert Einstein

The Vedic Times

The ancient India that gave the world the zero, Yoga and the original form of martial arts, traces its literary and scientific history to one of the oldest civilisations in the world. The knowledge and wisdom of this civilisation in varied fields were such that they were passed down from generation to generation through the oral tradition, with the earliest recorded written text, the *Rig Veda*, being written in textual form in Sanskrit in the Vedic times, somewhere between 2,000 and 500 BC. All the *Vedas* (Sanskrit word for knowledge) were written in the Vedic-Sanskrit language – the word Sanskrit itself means 'refined' or 'perfected' and the language is recognised worldwide as the oldest, most scientific and systematic in the world. In the Vedic world view, the processes in the sky, on the Earth and within the human mind are taken to be interconnected.[9] These sacred scriptures, made up of hymns, stanzas and *mantras*, not only trace the origins of the universe and expound the basic philosophy of the Trinity of God, but contain elaborate works on areas of applied knowledge in the fields of music, medicine, surgery, chemistry, military science and physics including particle physics.

While the *Rig Veda* contained references to cosmology, another one of the sacred texts is devoted exclusively to our solar system, the translation of which gives an insight to the amazingly advanced astrological knowledge acquired by ancient Indians. *Surya Siddhant* (Sanskrit for Sun solution), a textbook on archeo-astronomy compiled some time around 1,000 BC, for example, calculated the diameter of Mercury to be 3,008 km which actually is off-target of the currently known 3,032 km by only one percent[10] and calculated the distance between the Earth and the Moon to be 253,000 km as against 252,710 km that it actually is. The knowledge on applied astronomy in the Indian subcontinent, however, came about by the scholar and astronomer, Aryabhatta. Born in 476 AD, in Kerala,[11] while some accounts mention his birthplace in Bihar,[12] Aryabhatta is credited with, amongst others, the number zero,[13] the value of 'pi' which he calculated as 3.1416 as against the modern calculation of 3.14159, rotation of the nine planets in relation to the Sun, the Sun being the source of moonlight, causes of eclipse, discovery of the Earth being round and its circumference being 24,835 miles as against the modern calculation of 24,900 miles,[14] and his heliocentric theory of gravitation, which was one of the earliest in the world, preceding the European version by a thousand years.[15] Aryabhatta's main work is known as the *Aryabhattiya,* significant enough to be translated into Latin in the 13th century, enabling European mathematicians to work on Aryabhatta's method of calculating the areas of triangles, volumes of spheres, and square and cube roots. It was a rightful tribute to Aryabhatta when India's first indigenous satellite launched in 1975, was named after him.

The period up to the time the British landed on Indian shores, witnessed acquisition of an immense amount of knowledge in almost all branches of science. In the realm of particle physics, it was known that matter was made up of atoms (*anu*, in Sanskrit) and sub-atomic particles (*paramanu*) which on combining in different configurations determined different properties of substances in nature. It was known in the first century AD that it was light from an external source that illuminated the world around the human eye, that sound and light travelled in waves and light was much faster than sound. From the study of planetary motion evolved work on the laws of motion describing the curvilinear motion (*gaman*) and rotary motion (*bhraman*) in the 7th century. Bhaskaracharya, in the 12th century, took this work further and formulated the velocity, speed and time equation. Magnetism was posited in the 11th century, followed by the phenomena of electrostatic attraction.[16] Further work in the sciences lay by and large dormant during the Mughal occupation from 16th to the early 18th century and then in the age of the British colonial rule, and India emerged from this rather long period with its scientific history almost entirely forgotten.

It was in this context that Mark Twain was led to comment, "India is the cradle of the human race, the birthplace of human speech, the mother of history, the grandmother of legend, and the great grandmother of tradition. Our most valuable and most instructive materials in the history of man are treasured up in India only."

The Colonial Times

The Mughal reign was followed by two centuries of British presence as traders and then as rulers. While much of Europe was in the sway of the industrial revolution, its impact did not reach the Indian shores in the wake of the British ships that headed eastwards. Early scientific activity that did arrive on Indian shores was motivated by furthering the British interests in trade and commerce and later pressed into the service of empire building.[17] One of the very first activities undertaken by the British which were responsible for their success in domination over the vast Indian subcontinent was in getting to know the lay of the land. This led to the establishment of the Survey of India as early as in 1767, followed by institutions for astronomy and meteorology.

Indians began to see the role of science as a social reformer. As *The Hindu Patriot* wrote on April 6, 1854, "...the resources of the country will never be developed unless the children of the soil learn to develop them". Such were the sentiments that prompted Dr. Mahindra Lal Sircar to pioneer developing Indian enterprise amidst British rule. Armed with a vision to see Indians mastering what had already been discovered by the Europeans, and in due course of time, adding their own discoveries to those of the West,[18]

Sircar established the Indian Association for Cultivation of Science in Calcutta in 1876. The Indian Institute of Science at Bangalore followed in 1911, established by the entrepreneur Jamshedji N. Tata. The Nobel Prize awarded to C.V. Raman in 1930 bolstered domestic pride in Indian sciences, though World War II but economic hardships prevented any translation of the same into anything meaningful.[19] Pandit Nehru and Subhash Chandra Bose envisioned the role of science in free India as early as in the 1930s when the Indian National Congress started planning for science and industry, in anticipation of the independence that would allow them to implement such policies in an independent India.[20] The part played thereafter by notable scientists like C.V. Raman, S. Ramanujam, M.N. Saha and Satyendra N. Bose invited international accolades and was a major driving force for the Indian mindset and its self-esteem during its run-up to independence.

Independent India

The future belongs to science and those who make friends with science.

— Pandit Jawahar Lal Nehru

Pandit Nehru's vision of an independent India's industrialisation based on self-reliance, translated into the foundation of a firm science and technical infrastructure by way of public sector research enterprises, laboratories and industries. The early era of independent India witnessed establishment of, and then achieving tremendous strides by, the Council of Scientific and Industrial Research (CSIR), the Defence Research and Development Organisation (DRDO) and the Atomic Energy Commission. Notable amongst the architects of the country's Science and Technology (S&T) infrastructure were chemist/scientist Shanti Swarup Bhatnagar, nuclear physicist Homi J. Bhabha and the father of the Indian space programme, Vikram Sarabhai. To provide young engineers to further fuel the country's growth curve, Indian Institutes of Technology (IITs) were established, the first of which was as early as in 1951 in Kharagpur.

At the national level, the Department of Science and Technology was set up in 1971, which, besides being the apex body for policy formulation and coordination, also focussed on development of indigenous technologies in the emerging areas and futurology.[21] Imports of foreign technologies dropped significantly and indigenous Research and Development (R&D) rose, including in the private sector. Opening up of the largely state controlled economy, economic liberalisation and reforms in the early 1990s changed the Indian economic landscape and paved the way for the country to emerge as a nation that began to be counted amongst the emerging countries of the world. As a part of the economic renaissance, the Industrial Policy Statement of 1991 had among its many objectives, the aim of "injecting the desired level of technological dynamism into Indian Industry".[22] Indigenous R&D was given a fillip by strengthening the linkages between academia, industry, financial and R&D institutions. CSIR, as the premier industrial R&D organisation with an aim to rejuvenate the private sector, took on a more commercial orientation. By the end of 2009, CSIR had an impressive patent portfolio of 2,349 patents in India and 3,054 patents abroad while its 39 laboratories contributed to produce over 450 Ph.D. students in various disciplines.[23] The narrative on "India shining" would remain incomplete without a mention of the country's biggest asset – the Indian diaspora – a large English speaking skilled native workforce, including those armed with foreign degrees and experience, which contributed immensely to the economic growth.

It was this very skilled workforce, enabled by inflow of foreign investment that followed the deregulation policies, which stamped the Indian IT industry as a significant contributor to the domestic economy and placed the country on the global map. From a modest, 150 million US$ in 1990-91, the IT industry rose to a whopping US $ 88.1 billion in Financial Year (FY) 2011 and is expected to touch US $ 100 billion in 2012-13. This has created an impact not only on the nation's GDP but brought about a sea-change in the social milieu of the employable generation and the purchasing power of the burgeoning middle class. At the global level, India is now seen as a nation of innovative entrepreneurs with a massive reservoir of technical talent, capable of providing world class technological solutions and business services.

The Indian Space Saga

100th Milestone in India's Space Odyssey

In a landmark achievement in its ongoing space programme, India on September 9, 2012, launched its 100th mission into space. The Indian Space Research Organisation (ISRO) launched a commercial payload aboard its space workhorse, the PSLV–C21, the third such rocket to carry satellites purely for foreign customers. Two satellites were successfully placed in orbit – the French 712 kg optical remote sensing satellite, SPOT 6, and a 15 kg Japanese micro-satellite, the Proiteres. With this launch, ISRO has successfully launched 62 satellites, including 28 foreign, aboard 38 rockets.

The foundation of the space venture in the country was laid yet again based on the forward thinking policies enunciated by Pandit Nehru. The Indian Astronautical Society was formed within 10 years of independence, in 1957, with the aim to encourage domestic research on space topics. It took another 18 years to launch the first indigenous satellite, the Aryabhatta, in 1975, and then another five years to achieve an indigenous space launch. There has been no looking back since then in the success story of the Indian space venture. Though India has been nuclear-capable since 1974, the space programme remarkably had its origins for peaceful purposes. Its SLVs, unlike the other global space players, did not emerge from ballistic missiles but originated from civilian rockets and the space programme remains the most heavily supported civilian S&T programme in the country.[24]

The Motives Behind the Space Programme

There are some who question the relevance of space activities in a developing nation. To us, there is no ambiguity of purpose...we are convinced that if we are to play a meaningful role nationally, and in the community of nations, we must be second to none in the application of advanced technologies to the real problems of man and society.

— Dr. Vikram Sarabhai

Founder of the Indian Space Programme

The Cold War between the two superpowers was underway when an Indian diplomat and statesman V.K. Krishna Menon first used the term 'non-alignment', in 1953 at the United Nations. Pandit Nehru was not only an advocate of the Non-Aligned Movement (NAM) but also one of the founding fathers of the movement in 1961. Pandit Nehru's vision of non-alignment was to avoid dependency on either of the two blocs of power, and with seeking to occupy the middle ground, he sought to avoid being drawn into a possible third world war.[25] Aiming to distance itself from the arms race, India applied a self-imposed moratorium on developing ballistic missiles as a means of delivery of long range munitions. Dr. Vikram Sarabhai, the father of the Indian space programme, piloted the space programme towards applications that could functionally benefit the Indian masses in terms of education and health care, and bolster the nation's economy. Appreciating the enormous potential that space-based applications offered for enhancing the quality of people's lives, three areas were selected and energised for space development: communications, weather/climatology and remote sensing.[26]

Over the decades of execution of the Indian space programme, the earlier vision of Dr. Vikram Sarabhai of space-based applications was engineered for social needs, that encompassed capability-building in areas as far and wide as telemedicine, education, weather forecasting, remote sensing, agriculture, telecommunications, disaster management, environment monitoring and Direct-To-Home (DTH) television, aimed towards social upliftment and poverty eradication through a constellation of 20 in-orbit and operational satellites as of September 2012.

On the other hand, there are some who suggest that seen in purely economic terms, India's space assets appear to have cost more than the revenues that it generated.[27] Backed by a relatively modest[28] annual budget of around US $ 1.5 billion[29] for the year 2012-13, revenue-generation apart, the Indian space programme offers additional dividends by gaining tremendously in the realm of self-reliance, attaining a global standing that is technologically independent from other spacefaring nations. Over the years, the motive has undergone a paradigm shift by way of broadening into a vision that appears to have naturally gravitated from that of human and societal development to a stance that views space as not only enhancing the living conditions of its populace but also as a measure of global prestige.[30] Commencing in 1999, the Indian space initiative has entered yet another exclusive field – that of providing low cost launch services, rivalling the Russian and Chinese influence in this niche segment. Commerce is not necessarily the only motivator as such partnerships with global players offer tremendous foreign policy spin-offs in India's favour.

The year 2008 saw another glorious chapter being added to the Indian space-success narrative. The Indian vision for its space programme underwent a dramatic turnaround as the nation, instead of looking at the Earth from space, launched a payload that explored beyond-Earth space. The Chandrayaan 1 launched on October 22, 2008, was the country's first unmanned lunar probe. Its Moon impact probe made a controlled landing on the lunar surface 22 days later, making the country the fourth in the world to have its flag planted on the Moon.[31] It was this mission that made the world aware that water exists on the Moon. Of equal significance is that besides Indian payloads, the mission carried the most international lunar payloads ever, a result of the combined efforts of NASA, Bulgaria and 17 countries within the European Space Agency (ESA). This technological achievement, duly supplemented by an enormous amount of international cooperation paves the way for India's pivotal strides in the global space arena. With plans afoot for the Chandrayaan II and Mangalyaan to the Moon and for a probe to Mars respectively, India's motive would need to be beyond the scope of just acquiring 'international prestige'. In this context, it would be worthwhile to mull over a statement made in 2007 by the then Indian President and scientist-technologist, Dr. A. P. J. Abdul Kalam, in his speech (via teleconferencing) at a Boston University symposium. Commenting on the Earth's fossil fuels being depleted by the 21st century, Dr. Kalam advocated that a "space industrial revolution" will be necessary to exploit the high frontier's resources. He predicted that India will construct giant solar collectors in orbit and on the Moon, and mine Helium-3 — an incredibly rare fuel on Earth, but one whose unique atomic structure makes power generation from nuclear fusion potentially feasible — from the lunar surface where it is found in abundance. India's scramjet, the Reusable Launch Vehicle (RLV), Dr. Kalam asserted, will provide "low-cost, fully reusable space transportation" that has previously "denied mankind the benefit of space solar-power stations in geostationary and other orbits".[32] While such predictions could well chart a course for a futuristic space scenario, ISRO has rather pragmatically documented its "Space Vision 2025 and Beyond" that brings out the need for international synergy through collaboration and partnerships in space missions, to overcome the heavy costs involved to access space resources and explore inter-planetary space.

A Historical Perspective

Use of military rockets in the Indian battlefields as early as in the 1790s by Tipu Sultan, the King of Mysore, against the British East India Company, has already been brought out earlier in Chapter 2. In the case of space applications too, India did indeed have a history. Besides the theoretical context provided by Aryabhatta in the 6th century AD, astronomical observatories were established in various cities in the 1700s and 1800s that provided professional expertise that impressed European visitors.[33] The colonial subjugation kept at bay any further indigenous activity in matters of space and allied sciences and it was only after independence that Pandit Nehru brought back the focus in such a field, heralding the dawn of the Indian space age.

The year 1962 could well be the year that India took its first step towards space research by setting up the Indian National Committee for Space Research (INCOSPAR) which commenced setting up the Thumba Equatorial Rocket Launching Station (TERLS). Dr. Vikram Ambalal Sarabhai, a scientist and the father of the Indian space programme, while negating the notion of competing in a space race, argued in favour of "applying space technologies for the real problems of man and society".[34]

Dr. Vikram A. Sarabhai: The Pioneer of the Indian Space Programme

It would only be in the fitness of things to have an insight into the life of the visionary, Vikram Sarabhai, who pioneered and then charted the course of the Indian space saga. Born of a wealthy industrialist father and a social activist mother on August 12, 1919, in Ahmedabad, Gujarat, Sarabhai graduated from Cambridge in physics and mathematics in 1939. On return to India, he studied at the Indian Institute of Sciences under the Nobel laureate, Dr. C. V. Raman, and published his first scientific paper, "Time Distribution in Cosmic

Rays". After completing his PhD on cosmic ray physics at Cambridge in 1947, Sarabhai lost no time hurrying back to his homeland to almost immediately establish the Physical Research Laboratory (PRL) in Ahmedabad, at the age of just 28. He was the Director PRL from 1947-71, authoring 85 scientific papers and tutoring PhD students, 19 of whom did their doctorate directly under his supervision.

The launch of the Sputnik 1 by the Soviets in October 1957 was an event that galvanised Sarabhai into looking at the potential of satellite applications for societal development and brought to the government's attention the need to build an indigenous satellite. The government saw reason and INCOSPAR was set up in 1962 with Sarabhai as its first Chairman. While INCOSPAR's charter was to formulate the Indian space programme, in reality, Sarabhai continued to run the programme from his office at PRL in Ahmedabad.[35] An innovator with multi-faceted talents and interests, he set up the country's first market research organisation, the Operations Research Group, in 1961 and, along with a few spirited industrialists, played a central role in establishing the Indian Institute of Management, Ahmedabad, the same year. His academic and research oriented bent of mind manifested itself when he established a Community Science Centre in Ahmedabad in 1966, which today is known as the Vikram A. Sarabhai Community Science Centre. He spearheaded the establishment of the Indian Space Research Organisation (ISRO) in 1969, the establishment that was to fulfil Sarabhai's dream by launching India's first indigenous satellite in 1975, four years after Sarabhai's untimely death at the age of 52.

The visionary in him went beyond his calling. As an institution builder, he, along with his wife, founded the Darpana Academy of Performing Arts (1951) and contributed to setting up the Blind Men's Association (1954). These two were amongst 42 other establishments and institutions created by Sarabhai in fields as diverse as science, industry, technology, management and education.[36] All this, while he continued to run his family textile business and its network of companies from 1947 to 1966.

Vikram A. Sarabhai was conferred with the Padma Bhushan (1966) and Padma Vibhushan (1972) posthumously. After his death, the whole space establishment at Thiruvananthapuram has been renamed the Vikram Sarabhai

Space Centre. A crater on the Moon has been named Dr. Sarabhai Crater by the International Astronomical Union at Sydney in 1974.

Early Years of Rocketry

With INCOSPAR established in 1962, with Sarabhai as its Chairman, to spearhead the Indian space programme, building rockets was the obvious first step in any space venture, and ably supported by yet another doyen of Indian science, Dr. Homi Jehangir Bhabha, the father of the Indian nuclear programme, TERLS in Thiruvananthapuram was made operational the following year. The very first cooperative venture in the realm of rockets came about with permitting the Soviets to launch their scientific sounding rockets from TERLS which enabled the Indians to gain expertise in modern rocketry that aimed to eventually lead to indigenous capability in space launch vehicles.[37] It was largely due to India's non-aligned stance that it played the middle ground and reaped the benefits of collaborating with both, the USA and USSR. Indian scientists underwent training on sounding rockets at the NASA launch facility in Virginia. The first rocket launched from TERLS on November 21, 1963, was based on the design of the US Nike-Apache.[38]

A proposal by Sarabhai to run TERLS as an international launch facility was accepted by the UN in 1964. Under this arrangement amongst partners, NASA provided telemetry and ground support, the French supplied Centaur rockets and radar, while the USSR provided computers, ground testing facilities and a Mil helicopter.[39] With such an infrastructure in place, the first purely indigenous rocket, the 10 kg solid-fuelled Rohini 75 or RH 75 (75 indicating the diameter, in millimetres), was launched on November 20, 1967, from TERLS. Over time, Indian rockets became bigger,[40] went higher, carried better technology payloads and, most importantly, generated an enormous amount of hands-on experience which boosted confidence in testing advanced sub-systems and evaluating more powerful propellants. For example, the RH-125 was used to study the separation of rocket stages in flight and test remote-destruction in the case of a rocket going off-course. Similarly, three RH-125 rockets were clustered together to serve as a big booster.[41]

Building Space Infrastructure

Dr. Homi Jehangir Bhabha, the pioneer behind the nuclear programme and the then Chairman, Department of Atomic Energy (DAE), died in an air accident in 1966. Sarabhai, in addition to being Director, PRL and Chairman, INCOSPAR, was tasked to assume Bhabha's responsibility. ISRO was established in 1969 under the DAE,[42] for space technology development. Buoyed by the success achieved in sounding rockets, ISRO set about on development of an indigenous SLV, to enable India to have independent access to space. To provide focus to the space programme, ISRO, with Prof. Satish Dhawan as its Director after Sarabhai's death in 1971, was delinked from DAE and placed under a newly created Department of Space (DOS) in 1972, with a strategic plan to place a 40 kg satellite into a LEO. Recognising the difference between the needs of a developing country and the route taken into space by established space players, Sarabhai favoured 'leap-frogging' the process of development by acquiring advanced technologies and then developing them to suit domestic requirements.[43]

By this time, the facility at TERLS had grown in size and scope. To facilitate design, development and construction of rockets, payloads and instruments and to promote research in space science and technology, the Space Science and Technology Centre (SSTC) was set up in Thiruvananthapuram in 1969. The Rocket Propellant Plant was commissioned the same year, right next to Thumba to make solid-propellants for the rockets.[44] The three launch pads at TERLS, limited to launching rockets of diameter of only up to 0.56 m, were obviously inadequate to realise the dream of launching bigger and heavier SLVs. A launch site at Sriharikota Range (SHAR), an island of over 180 sq km on the east coast off Andhra Pradesh, was selected and a Rocket Fabrication Facility (RFF) to make rocket hardware was set up in 1970 at SHAR. SHAR became operational in October 1971 and today has two launch pads that generate two to three launches per year.

In the field of Earth stations for exchanging data with satellites, the Ahmedabad Earth Station was the first one to be set up, in 1967. Built indigenously on the insistence of Sarabhai, as against the desire of the Minister of Communications who was about to ink a deal with the US company RCA for its construction, the station was set up in a record 87 days.[45] It was built to receive data from the American Applications Technology Satellite (ATF) in a venture with the USA, for imparting education to rural India under a project initiated by Sarabhai, called SITE (Satellite Instructional Television Experiment).

Space Launch Vehicles Take Flight

The Project Director appointed for the indigenous SLV programme was Dr. A. P. J. Abdul Kalam, who later became the President of the nation. Development of the four-stage solid-fuelled SLV-3 commenced in 1973, with plans of a 17 ton launch mass, 13 tons of which was fuel. The 22.5 m tall rocket with one metre diameter was tiny compared to those being launched in that era, but was a huge advance for the country as the largest sounding rocket launched so far was the Rohini 560, with a diameter of 0.56 m. The SLV-3 had 44 main systems totalling 10,000 components, 85 per cent of which were indigenous[46] with 46 organisations in the public and private sectors involved in the project. The SLV-3's maiden launch from SHAR on August 10, 1979, failed five minutes into the flight due to the failure of the second stage.[47] The SLV-3's second launch on July 18, 1980, was a success and its payload, the 40 kg remote sensing satellite Rohini, was successfully put into an elliptical orbit, heralding India's entry into to a select group of launch-capable spacefaring nations, the seventh in the world to do so.

Building upon this heady success, India's ambition now was to place heavier satellites into orbit, enhancing the capability from 40 kg to a one-ton satellite aboard a Polar SLV (PSLV). This too would necessitate a significant jump in technology and the solution was found in an incremental enhancement in development. A low cost 'augmented' version of the SLV (ASLV) was the interim measure: a five-stage, solid fuelled rocket designed to place a 150 kg satellite into a 400 km circular orbit.[48] The 23 m tall but 39 ton rocket enabled heavier payloads, so the new satellites were called 'stretched' Rohini satellite series, or SROSS. The first two launches in 1987 and 1988 failed but the third launch of the ASLV in May 1992 successfully placed a SROSS C satellite in a near-intended orbit. The fourth and final launch was the most successful, orbiting the 113 kg satellite, SROSS C2, in May 1994.

While the SLV-3 and the ASLV were designed as experimental launch programmes, it was the PSLV that conferred the rites of passage into the operational space launcher technology and earned for India an entry into the 'big launcher league'. To place in orbit larger satellites of one ton class that was envisaged for the series of Indian Remote Sensing (IRS) satellites, the need was to exploit the advantages of power and control that liquid-fuelled rockets offered. The 44 m tall and 275 ton PSLV was almost ten times bigger and heavier than the ASLV and had a combination of alternate solid-liquid stages in its four stages with strap-on boosters.[49] Following the pattern of the ASLV launches, the first launch of the PSLV in September 1993 failed, losing its payload, the IRS-1E satellite. The second, in October 1994, was perfect and placed the 870 kg IRS P2 satellite into an 825 km orbit. This feat allowed the country to break away from its reliance on Russian launchers for orbiting the IRS series of satellites and in addition made the domestic launch services commercially available to foreign partners. With the launch of a PSLV-C25 in November 2013 which placed the Mangalyaan mission enroute to Mars, PSLVs have established an enviable track record of 24 consecutive successes in 25 launches and continue to launch and orbit satellites in LEO.

The most notable amongst satellites launched into LEO remain the series of IRS satellites that have enabled the nation to have the largest constellation of remote sensing satellites in the world, with some of the newer ones providing sub-metre resolution. Other notable and successfully orbited payloads and missions launched aboard the PSLVs include the first commercial launch in May 1999 that orbited three satellites, including one each of Korea and Germany; the Technology Experimental Satellite (TES) launched in October 2001 that provides imagery of one metre resolution, making the country the only one other than the USA to possess such capability[50] at that time; Kalpana, the first indigenous meteorological satellite in September 2002; HAMSAT, a 42.5 kg micro-satellite for amateur radio services in May 2005; SRE 1, the space capsule recovery experiment that was deorbited and recovered after 12 days in January 2007; mission PSLV C9 that placed ten satellites into orbit simultaneously in April 2008 and made the nation the first in the world to do so; Chandrayaan 1, the unmanned lunar probe in October 2008; RISAT 1, the first indigenous all weather radar imaging satellite weighing 1,858 kg in April 2012, and Mars orbiter, Mangalyaan, in November 2013.

The Chandrayaan 1 mission to the Moon employed a variant of the PSLV named PSLV-XL that had more powerful stretched strap-on boosters, taking the weight of launch mass to a mammoth 320 tons. The Kalpana mission was unique in another aspect. Though the PSLV was designed to launch satellites into polar orbits typically at 900 km, ISRO added half a ton extra propellant in the fourth stage and stretched the polar orbit into a geosynchronous orbit, and placed a 1,060 kg meteorological satellite, Metsat, into the GEO slot of the aging INSAT 1D, in September 2002. Metsat was later renamed Kalpana in memory of the Indian-American who perished along with the crew in the Space Shuttle Columbia disaster.

To gainfully exploit the complete spectrum of space applications, one essential segment was of having dedicated satellites operating in geostationary orbits: satellites that would orbit once every 24 hours and would appear to be permanently 'parked' 36,000 km over a geographical point in the Indian subcontinent. The Indian National Satellite System (INSAT) planned as early as in 1976, aimed to cater to this emerging national need by providing communications and weather monitoring, and later on for navigation and TV services. Launching satellites into GEO required yet another quantum leap in launch technology. While the first nine INSAT launches came about aboard foreign launchers, for economic reasons and, more importantly, the nationalistic desire for self-reliance, it was in 1987 that India felt the need for indigenous launchers for orbiting its two-ton satellites in GEO, a need that gave birth to the Geostationary Space Launch Vehicle (GSLV) programme. The design of the 49 m tall, 414 ton, three-stage GSLV caters for the already proven solid-fuelled first stage of the PSLV as its first stage[51] and for strap-on boosters, relies on four liquid-fuelled rockets instead of the conventional solid-fuelled strap-ons. The second stage is the same proven Vikas[52] engine that formed the second stage of the PSLV that sent the Chandrayaan 1 to the Moon. The development and operationalising of the third stage, the liquid-fuelled cryogenic engine,[53] turned out to be an altogether different story, both technologically and politically.

In 1991, the USSR and India entered into an agreement wherein India would buy cryogenic engines and related production technology. Having apprehensions that such technology would violate the provisions of the Missile Technology Control Regime (MTCR) by being used for developing long range ballistic missiles,[54] the USA applied sanctions on this deal in 1992. The reason for such apprehensions was not necessarily violation of the MTCR – as cryogenic fuels take a long time to prepare on the ground and have no utility for military rockets that need to be launched at shorter notice – but rather commercial.[55] Prior to the deal with the Soviets, India had approached Japan for its LE-5 cryogenic engine which did not fructify. Getting wind of this, the American General Dynamics Corporation offered the cryogenic engine and its technology transfer. India did not express interest as the cost was found to be prohibitive. Some are also of the view that the American intervention in the USSR-India deal was to keep India out of the commercial space launcher market. Subsequent events that unfolded, delayed the acquisition of cryogenic technology by five to ten years[56] but in the larger perspective, strengthened the Indian resolve to set course towards indigenous development of the cryogenic engine.

Post the break-up of the Soviet Union, the need for hard cash made Russia initially reject the US request for sanctions. The American financial aid package to Russia in 1993,[57] however, turned the tide and Russia withheld the Indian deal, agreeing only to the sale of seven completed cryogenic engines, the KVD-1.[58]

The first successful launch of a GSLV fitted with the Russian cryogenic engine placed an educational satellite, the 1.5 ton G-SAT 1, in a near-GEO in April 2001. Of the balance six Russian rockets, five more were expended in GSLV launches between May 2003 and December 2010, three of which were successful. The nation's first indigenous cryogenic engine fitted on the GSLV, the GSLV Mark II, was launched in April 2010 but crashed into the sea five minutes into its flight, due to underperformance of the crucial third stage's cryogenic engine. With only one KVD-1 engine in its hand, it is for ISRO now to draw lessons from the failure of the first indigenous cryogenic engine and face up to the challenge of bringing to frution the ambitions of self-reliance by honing the indigenous content in launching geostationary satellites. While the saga of Indian space launchers has been creditable, notwithstanding a few failures that are inherent in any major path-breaking technological programme, the space launcher industry, at the very least, should aim to catch up with the domestic satellite industry whose progress and reach has been even more remarkable.

Towards this end a programme for an improved version, nay, a technological successor of the GSLV has already been approved by the government in 2002. The GSLV Mk III aims to place a five ton class satellite into GEO, besides providing a multi-mission capability in LEO, polar and other intermediate circular orbits, with an aim to make the nation fully self-reliant in launching heavy satellites into GEO as also to enable the country to be a competitive player in the multi-billion dollar commercial space launch market. This three-stage GSLV will be shorter, at 42.4 m, but mightier and more powerful, with a launch weight of 630 tons. The first stage comprises two identical large solid-fuelled boosters that will be strapped on to the second stage, the restartable liquid-fuelled double Vikas engines. The third stage will be the powerful cryogenic stage[59] with 25 tons of propellant, double that of the KVD-1. It is apparent that the design not only relies on the currently available capabilities with ISRO, but armed with a cryogenic engine, would be favourably placed for future augmentation.[60]

Space-based Applications for the Society

> *Science alone can solve the problems of hunger and poverty, insanitation and illiteracy, of superstition and deadening customs and traditions, of vast resources running to waste, of a country inhabited by starving people.*
>
> — Pandit Jawahar Lal Nehru

Capability Enhancement in Space

In the 1960s, when satellite-based applications were still in the experimental stages even in the USA and USSR, Dr. Sarabhai was quick to understand the implications of such applications in support of, and to supplement, ground-based systems. While charting the course for developing indigenous satellites, Sarabhai realised the need to tap into existing space assets by initially banking on foreign satellites for proving applications, building the requisite ground support infrastructure and then innovating to indigenise.

The very first venture initiated by Sarabhai was the Satellite Instructional

Television Experiment (SITE) aimed at education of the large rural-based masses. While the capital of the country had TV as early as in the 1950s, its reach was limited to 30 km and it had no national coverage. All-India TV proposed a national grid, with ground masts in 17 major cities and spreading the infrastructure outwards. The visionary in Sarabhai took the opposite approach, that of connecting remote villages first through geosynchronous satellites. An agreement signed with NASA in 1969 saw the American satellite, the American Technology Satellite-F (later called ATS-6) moved into a GEO slot over India and it broadcast education and health related programmes[61] over the Indian landmass in 1975-76, under the SITE initiative. SITE was called "the largest sociological experiment in history".[62]

In the field of remote sensing, so intrinsic to the domestic need of mapping the natural resources and landmass of the seventh largest country in the world, India became one of the earliest users to capitalise on the data received from the world's very first civilian remote sensing satellite, the American satellite Landsat that was launched in 1972.

With these two early experiments, India had forayed into, and experienced first-hand, space-based applications that emerged from both the distinctly different orbits – education and communications from the ATS-6 satellite in GEO, and remote sensing from the Landsat satellite in LEO. The ground station network had commenced building up and continued expanding. ISRO's establishment came about in 1969 and its strategic charter commenced with the plan to launch a 40 kg satellite into LEO. Dr. Sarabhai presented a plan on the Indian National Satellite System (INSAT) in 1970 that involved procuring the first set of satellites from foreign partners and building the subsequent series domestically.[63] Within the domestic satellite programme, ISRO commenced its work in essentially two fields. The first was in the field of remote sensing, the Indian Remote Sensing Satellites (IRS) system and the other in the realm of geosynchronous satellites, by way of the INSAT system for communications, TV broadcasting and meteorological services. INSAT aimed at another first in the world – to harness the advantages of communications (in GEO) and Earth observation (in LEO) from a common satellite platform in GEO, an approach more complex but 40 per cent cheaper in the long run.[64]

Saga of the Indigenous Satellites

It is no wonder that the first indigenous satellite, the Aryabhatta, bore fruit much before the indigenous launcher, as is the narrative the world over where over 60 countries have developed their own satellites but rely on foreign space launchers to launch and orbit them, a capability possessed by only 11 countries in the world. The country's first indigenous satellite, the Aryabhatta, a 360 kg scientific satellite for study of the Earth's atmosphere launched aboard a Soviet Cosmos 3M rocket in 1975 on a 594 km circular orbit, had a designed life of 17 years, but operationally lasted only five years.[65] The first step towards remote sensing was the 442 kg satellite Bhaskara-I launched in June 1979 from the Soviet Union which provided one km resolution photographs. It sent 10 pictures a day and provided valuable information on snow melting in the Himalayas, river flooding in North India, desertification in Rajasthan, rainfall in the coastal belts and mineral resources in Gujarat. The Bhaskara-II launched in 1981 returned the most useful data in its ten years life providing information on agriculture, vegetation and weather, and assisting in making maps of Bengal. All the three satellites were built by Hindustan Aeronautics Ltd (HAL), with the Soviet Union providing solar cells, batteries and thermal paints.[66]

Came the time to test the indigenous space launch capability, as the SLV programme had progressed on course, it was obvious that the first step to orbit the satellite would be small and tentative. The satellite Rohini that was launched by the first indigenous SLV in SHAR in July 1980 was a mere 35 kg mass, though equipped with a TV camera of one km resolution. While the SLV was to prove the indigenous launch capability, Rohini's key objective was to prove the Indian made solar cells. Both stood proven. Upbeat and assured with this success, the Rohini 2's launch in May 1981 was covered live on TV and radio. The Rohini 3, launched two years later, performed even better by beaming back 5,000 images, picking out water, vegetation, snow and clouds.

The IRS System Takes Shape

The indigenous satellite-building capability stood proven with the successes achieved in the experimental satellites of the Bhaskara and Rohini classes.

IRS satellites were designed to be one ton weight satellites, revisit or cross the same point on the Earth every 21 days and equipped with three linear imaging self-scanning high resolution sensors, to carry out systemic mapping of the Earth's surface. The first IRS, IRS-1A was ready but the SLV/PSLV had not kept pace and the IRS-1A was launched in March 1988 on a Rostok rocket. The 975 kg satellite was put in a 904 km polar Sun-synchronous orbit of 103 minutes (22 days revisit) providing a resolution of 36.25 m. It operated for over eight years,[67] returning 400,000 images that enabled comprehensive mapping of forests, salt land, water and wasteland. IRS-1B, specifically designed to forecast crop yields of tea and coffee, was launched in August 1991, again from the Soviet Union.[68]

The first in the series of IRS satellites to be launched by an indigenous PSLV was the 804 kg IRS-P2 (P for polar) in October 1994. The satellite-capability envelope thereafter was raised and the country moved to its second generation IRS satellites, the first of which and the only one thereafter, was to be launched aboard a foreign launcher. India orbited the 1,250 kg IRS-1C in December 1995 aboard the Russian Molniya rocket and the satellite's operation was unique in two ways. Its panchromatic camera with a 6 m resolution provided a significant boost in the accuracy of remote sensing, making it the most advanced remote sensing satellite in the world at that time.[69] Such an output enabled digital mapping of the whole country in a 1:125,000 scale and coupled with the output from the infrared-band sensor, fed into the early years of mapping of the domestic Geographical Information System (GIS). The other ground breaking activity linked to the IRS-1C was India's entry as an exporter at the global level. India permitted a franchise to EOSAT, an American company, to sell this imagery to the global market while the Japanese space agency, NASDA, bought the IRS data for its own remote sensing centre.

The PSLV programme stabilised, grew from strength to strength and the days of dependency on foreign launchers for the IRS satellites were over. Capability-enhancements became ever more specialised as satellites dedicated for different roles began to be launched into LEO. The all encompassing IRS series gave way to a variety of satellites that are designed and operated exclusively for individual facets of remote sensing, with satellites being named as per their area of application. The OceanSat (IRS P4) launched in May 1999, carried payloads exclusively for oceanographic studies and had instruments that could see up to 200 m depth. The Technological Experimental Satellite (TES) launched in October 2001 to experiment and validate new technologies in spacecraft and payloads, provides a one metre resolution that made the country the only one other than the USA to possess such capability at the time, raising apprehensions in some who saw the implications of such a capability in matters military.[70] Metsat in September 2002, later renamed Kalpana, orbiting in GEO, was specifically for meteorological purposes. It carried out hourly scans in the visible, infrared and water vapour bands, collected data from weather stations across the country and relayed them to the meteorological centre at New Delhi. The Cartosat series of satellites operate for digital mapping, town planning, and road and canal building. The highly agile and steerable Cartosat 2B launched in July 2010 provides a resolution better than one metre.[71] The technological graph gained a notch with the launch of a radar imaging satellite, the RISAT-2 in 2009. This was the country's first observation satellite with an all-weather radar imaging application for disaster monitoring. It has applications in the field of radar-based tracking of ships, a special need that arose in November 2008 during the Mumbai blasts, necessitating an outright buy from Israel within five months of the catastrophe.[72] The state-of-the-art RISAT 1 took this application further. Launched in April 2012, the 1,858 kg indigenous satellite, the heaviest launched by the PSLV into LEO, carries a microwave synthetic aperture radar operating in the C band, that enables day and night imaging operations in all weather conditions.

With over a dozen satellites currently in operation, the IRS system is the largest civilian remote sensing satellite constellation in the world providing imageries in a variety of spatial resolutions, spectral bands and swaths.[73] Resolution available today is that of less than one metre. The data is used for wide-ranging applications covering agriculture, water resources, urban development, mineral prospecting, environment, forestry, drought and flood forecasting, ocean resources, conservation of wildlife and disaster management, activities across a broad spectrum that are contributing immensely to India's economic, environmental and social development.

The INSAT is in Place

Dr Sarabhai's presentation in 1970 on the envisaged INSAT system operating in GEO, found favour with the powers-that-were and approval of the programme was accorded in 1977. GEO slots were registered two years later with the International Telecommunication Union (ITU) for orbiting INSAT satellites in geosynchronous orbits. The INSAT programme involved the Ministries of Telecommunications, All India Radio, and Departments of Space and of Meteorology. With the rather unique aim of providing applications in a dual role, that of communications and weather systems, an agreement was finalised with the American Ford Aerospace Corporation, now Space Systems/Lorel, in 1978 to build the first of the INSAT-1 satellites[74] and with NASA for launching them on their Space Shuttle. Events unfolded thereafter in a manner that allowed India to collaborate with the European Space Agency (ESA) too in the field of space launch. The first four flights of ESA's Ariane launcher were development flights and ESA offered free space to ISRO – free, as these were test flights and there was an inbuilt element of failure. The Indian response was prompt. The 630 kg[75] satellite, Ariane Plane Passenger Payload Experiment (APPLE), built by HAL, took 36 months to be designed, built, tested and shipped to Kourou, French Guyana, from where it was launched aboard Ariane V3, the third flight, in July 1981. This was the period when the indigenous satellites Aryabhatta, Bhaskar 1 and Rohini had already been launched. The challenge in building APPLE lay in the tight time schedule and, more importantly, in building the onboard apogee kick motor that would be required to manoeuvre and position the satellite in its orbital slot. Innovation was quickly adopted and the fourth stage of the SLV came in handy as the apogee kick motor for APPLE. Its two-year mission life provided for various communication experiments, including TV relays and radio networking. Even more absolute was the Indian hands-on experience and confidence in the GEO-based space relay platform that would pave the way for the indigenous INSAT 2 series of satellites.

The Space Shuttle's entry into service suffered numerous delays and necessitated launching the INSAT 1A by the American Delta rocket in April 1982. The satellite faced numerous problems in the deployment phase and was finally decommissioned after four months. The INSAT 1B, launched aboard the Space Shuttle Challenger in August 1983, was operational for almost 10 years, providing TV coverage to 70 per cent of the Indian population and returning 36,000 images[76] of the subcontinent. Post the Space Shuttle Challenger disaster, INSAT 1C switched to being launched by ESA's Ariane in 1988. Failure in the power system resulted in abandoning the satellite within four months. It was the Delta rocket again that launched the last of the first generation INSAT 1 satellites in 1990. The most successful of the four, the INSAT 1D continues to beam back quality images as also provides TV services in its footprint.

The indigenous INSATs commenced with the second-generation INSAT 2 series satellites that, however, still required foreign launchers. These were designed to be near 2 ton weight at lift-off, more than one and a half times that of the INSAT 1 series. Five satellites in this series were launched in the period 1992-99, all aboard ESA's Ariane rocket. INSAT 2A and 2B were near identical 1,906 kg multi-purpose satellites that provided communications (C, extended C and S bands), meteorology and satellite-based search and rescue services. While the Ariane rocket powered both to the geostationary transfer orbit, the final manoeuvring into GEO slots by firing the liquid apogee motor was controlled by the master facility at Hassan, Karnataka. The INSAT 2C and 2D satellites were purely for communications and, in addition, had transponders in the Ku band. The INSAT 2D became inoperable four months after launch while the INSAT 2C provided coverage to remote parts of the Northeast, Andaman and Nicobar Islands and as a yet another first, with enlarged C band transponders, reached out to countries beyond the subcontinent. The higher frequency Ku band enabled higher-density traffic while transmission began being received by small rooftop antennae. The last in this series, the multi-purpose INSAT 2E weighed 2.55 tons and besides the standard fare in payload, carried, for the first time, a high-resolution sensor for the water vapour band and a Charged Coupled Device (CCD) camera that provided one km spatial resolution. It collected and retransmitted data from 100 weather stations in remote areas. The performance of this satellite was such that a European consortium, INTELSAT, leased 11 transponders for 10 years at Euro 188 million. With the INSAT 2 system in place, 90 per cent of the population was covered by space-based TV. By the turn of the century, the INSAT system was extensive enough to provide video teleconferencing and education for farmers and rural development workers through 4,700 voice circuits via 430 Earth stations and 1,200 terminals. Fishermen in the coastal areas were being guided to fishing grounds by the space platforms. The ground stations for INSAT based search and rescue applications set up at Bangalore and Lucknow not only serve distress calls to the mainland but act as relay stations for countries as far as Indonesia in the east to Tanzania in the west.[77]

The INSAT 3 series satellites were driven by increased domestic demand for communication channels. Four of them were launched in the period 2000-03, all on Ariane, with the aim of placing more transponders in space as also to step into the GEO slot and fulfil the role carried out by the aging INSAT 1 and 2 series satellites. All four are multi-purpose satellites in the 2-3 ton class, barring the INSAT 3E which is purely a communications satellite, launched to augment the services being provided by the preceding INSAT satellites. The fifth in this series, the INSAT 3D, is purely a weather satellite planned for launch in 2013.

The need for the INSAT 4 series satellites arose essentially to build upon the services that were already in use, primarily in the rapidly growing fields of telemedicine and DTH TV. Four INSAT 4 satellites were launched between 2005 and 2007, of which one failed to reach its orbit. The INSAT 4A enabled ISRO to connect 33 specialised hospitals to 132 remote, rural or district hospitals in its telemedicine network in 2006,[78] while the figure today stands at 80 hospitals connected to 306 remote/rural hospitals and 16 mobile telemedicine units.[79] The INSAT 4B is an exclusive communications satellite while the INSAT 4CR provides DTH, video picture transmission and digital news gathering.

Widening the Spectrum of Space Applications

The IRS and INSAT systems might have been the initial key drivers of the space-based applications that were envisaged for domestic development, but that did not prevent the country from foraying into other fields related to space technology. The partnership forged between India and the Soviet

Union in the early years of Indian satellites being launched by the Soviets gave an additional benefit. Sqn Ldr Rakesh Sharma, a fighter pilot in the Indian Air Force, became the first Indian and the 139th human to reach space. Rakesh Sharma, along with two cosmonauts, was launched aboard the Soyuz T-11 on April 3, 1984, for a week-long stay on the Salyut space station. This could have been considered a singular, though remarkable, feat but for the fact that it gave birth to the idea of an indigenous manned space flight in the years to come. True enough, the Space Recovery Experiment (SRE) was launched in 2007 aboard the PSLV and the spacecraft safely deorbited and recovered back to the Earth 12 days later. This was the natural first step to prove the desired capability of safely recovering space travellers back.

While the INSAT series of satellites provided services in communications, weather and Earth observation, it was in the field of rural development that a new range of geosynchronous satellites was conceived. The Department of Space and the Ministry of Rural Development partnered in 2000 to introduce the GRAMSAT (*Gram* is the Hindi word for village) programme aimed to transmit TV, CD-quality sound, data and the Internet over the communications network connecting state capitals to districts and rural blocks. The terrestrial grids were planned to be connected via communications satellites. Thus, was born the GSAT series of satellites that were based on the INSAT 2 design with six powerful transponders on a national beam for educational TV and two spot beams to transmit sound in four different languages.[80] Commencing with the launch of the GSAT-1 aboard the GSLV in 2001, a total of six GSATs has been launched, including the last one, the GSAT-10 aboard Ariane in September 2012. Besides catering to the need of advanced communications for the GRAMSAT initiative, this constellation of satellites has equally importantly, also augmented the communications services of the INSATs by adding transponders in space. GSATs 8 and 10 have in addition carried payloads for the indigenous satellite-based navigation system, the GPS Aided GEO Augmented Navigation (GAGAN).[81] GSAT 7 was launched into GEO on August 30, 2013 aboard the Arianne. While this was the 71st Indian satellite to be orbited, its uniqueness lies in the fact that the GSAT 7 is the first to be publicly championed as a pure military satellite – one which is dedicated for the communications needs of the Navy.

Indian geostationary satellites, that is, the complete series of INSATs and GSATs, occupy GEO slots of 55, 74, 83 and 93.5 degrees east longitudes,[82] some of which have three or more satellites parked together, including a few obsolete ones.

To encourage space sciences among students of the domestic universities, ISRO has invited partnerships for building satellites and then launched them on home-grown PSLVs. The first of these was the 40 kg mini-satellite AnuSat, designed by students of Anna University. The satellite has a message store and forward payload and was launched in 2009. StudSat, a one kg pico-satellite built by students of seven engineering colleges of Karnataka and Andhra Pradesh, followed a year later and provides 90 m imagery of the Earth. YouthSat is a product of a collaborative venture between students of India and Russia that was launched in 2011 for the study of the atmosphere. Similarly, SRM University built the 10.9 kg SRMSat that was launched in 2011 for atmospheric study while IIT, Kanpur, built a 3 kg nano-satellite Jugnu (Hindi for firefly) to prove the student-developed image sensor in the near IR region.

Ground-Based Support Infrastructure

While ISRO is headquartered at Bangalore, there is almost a score of functional space establishments under ISRO spread across the country. The front face of ISRO, of course, remains the space complex at Satish Dhawan Space Centre at Sriharikota Range (SHAR) that has two launch pads which launch the entire range of PSLVs and GSLVs,[83] launching satellites into LEO, GEO and polar orbits, and also provides the complete gamut of infrastructure from assembly, testing, fuelling to launching of the rockets. Some of the other significant establishments include:

- The Vikram Sarabhai Space Centre (VSSC) at Thiruvananthapuram, the flagship of ISRO, is the apex research and development centre for satellite launch vehicles and associated technologies.
- Development of payloads for communication, meteorological and remote sensing satellites is undertaken at the Space Applications Centre (SAC), Ahmedabad.

- Telemetry and tracking of satellites is controlled from the ISRO Telemetry, Tracking and Command Network (ISTRAC) at Bangalore which has a network of ground stations at Bangalore, Lucknow, Sriharikota, Port Blair and Thiruvananthapuram in India besides stations in Mauritius, Bearslake (Russia), Brunei and Indonesia. ISTRAC also caters for support to its deep space missions including cross-support to similar missions of foreign partners.[84]
- Monitoring and control of all geostationary satellites is done by the Master Control Facility (MCF) at Hassan, Karnataka, and Bhopal, Madhya Pradesh.
- In the field of remote sensing, the National Remote Sensing Agency (NRSA) at Hyderabad is responsible for satellite data acquisition and processing, data dissemination and decision support for disaster management. Data, both digital and photographic, is then distributed to regional centres across the Indian landmass.
- To fill up existing vacancies of scientists and engineers in ISRO, an Indian Institute of Space Science and Technology (IIST) at Thiruvananthapuram was sponsored by ISRO, offering undergraduate, postgraduate and doctoral programmes focussing on space science, technology and associated applications commencing 2007.[85]

ISRO's Commercial Ventures

Reforms in the national economic policies in the early 1990s led to increased privatisation and the space programme benefitted by privatising technologies developed by ISRO.[86] This paved the way for establishing the Antrix (Sanskrit for sky/space) Corporation Limited in 1992, a marketing agency of ISRO under DOS with access to resources of DOS as well as the Indian space industries. Through the years of the space narrative, there developed a robust partnership between ISRO and other industries, both public and private. Over 500 companies participate in the space effort,[87] with almost 60 per cent of launch vehicle cost flowing to the industry.[88]

While Antrix's objective remains to facilitate development of space related capabilities in the domestic industry, of late it has also engaged in providing space products and services to international customers worldwide[89] achieving a turnover of Rs 884 crore in the year 2009-10.[90] In terms of sales, not only is the Indian space workhorse, the PSLV, the choice of many countries in launching their satellites in LEO, the collaborative approach has enabled many countries to build their satellites with ISRO's help. Revenue is also generated from leasing out lines on the INSAT satellites, providing Earth imagery and offering space-related consultancy services. Due to the relative low cost and robust reliability of the PSLV, the most commercially viable venture remains the space launch sector. Commencing commercial operations with the launch of the German TUBsat and South Korea's Kitsat3 satellites aboard the PSLV C2 in May 1999, ISRO has launched and orbited the satellites of many of its foreign partners that include Belgium, Argentina, Indonesia, Italy, Israel, Denmark, Japan, Canada, Netherlands, Switzerland, Turkey, Algeria and France. While most of these foreign-built satellites were successfully launched aboard the reliable PSLV, ISRO's credibility in building heavy geosynchronous satellites for foreign buyers got a major fillip when ISRO built satellites for EADS Astrium.[91] ISRO-built state-of-the-art communications satellites, the 3,463 kg W2M, and the 2,541 kg HYLAS (highly adaptable satellite) were successfully launched into GEO orbits by Ariane rockets in 2008 and 2010 respectively. Not only is the ISRO's satellite-delivery time of 28 months lower than the average global time schedule of 30 months, the cost of manufacturing a two-ton satellite at US$ 40-60 million is also lower, while the launch on a PSLV is at US$ 20 million.[92] Continuing the field of internationalising domestic space-based applications, imagery from the Indian remote sensing satellites has found users in Australia, Africa, Europe, South Korea, China, Japan, Nepal, USA, Turkey, Vietnam and Sri Lanka, based on a network of 22 international ground stations and 20 resellers worldwide.

Going beyond individual services that ISRO offers commercially, at the government level, there have been Memorandums of Understanding signed with over two dozen foreign governments and space agencies across the world wherein ISRO will provide expertise and services for helping other developing countries in the applications of space technology.

The Years Ahead

Twenty years from now, when space travel is likely to become mundane like airline travel today, we don't want to be buying travel tickets on other people's space vehicles.

— Dr G. Madhavan Nair
Chairman, ISRO, May 2008

Having firmly established a place for itself in the new high-ground, that of space, India looks at the years ahead as an era of opportunities. Showcasing its vision from practical space applications for 'service to humanity', to a path of national pride and thence onto consolidating its stature as that of a responsible spacefaring nation, India today understands the need to look at the rapidly developing global space order and adapt its space strategy in a manner that is intrinsic to its interests, addressing simultaneously the platforms at the domestic, regional and international levels. In the all-encompassing global geopolitical milieu, the nationalistic focus on self-reliance is not likely to be lost sight of. Towards this end, the immediate aim would need to be in the launch sector, by gaining success in the critical cryogenic technology that has so far eluded the indigenous GSLV Mk II and then following it up with its technological successor, the GSLV Mk III that would place a five-ton satellite in GEO. To increase the number of launches per year, the DoS has approved a ₹ 350 crore project to add one more rocket assembly line at the second launch pad at SHAR, a project that awaits the government's nod and, in addition, is contemplating plans for a third launch pad at the space port.[93] Self-reliance in this core capability of space launches, besides the issues of pride and cost savings, would offer tremendous benefits by way of geopolitical spinoffs.

In the field of satellites, ISRO announced on October 8, 2012, that it plans for 58 space missions in the next five years to place 33 satellites in orbit.[94] Besides just the numbers of satellites, it would be more prudent to look at the applications that are planned to be fielded. The INSAT and GSAT series to be launched in the years to come, would continue to add transponders in space to meet the burgeoning need of additional bandwidth for domestic communications as well cater to those of foreign partners. Equally important are service-based systems that are planned for the midterm. One of the notable among them is the GPS Aided Geo-Augmented Navigation (GAGAN) system, a satellite-based regional navigation service that is being fielded by the Airports Authority of India (AAI) and ISRO that aims to provide air traffic management and navigation over the Indian air space, making India the fourth in the world to do so. While three ISRO-built geostationary satellites will provide tracking facilities from space, the US-based defence technology company Raytheon has been contracted to set up 15 ground stations in support of the system. GSATs-8 and -10 with GAGAN payloads are already in their GEO slots and provide better than 7 m accuracy[95] while the third satellite to complete the constellation, the GSAT 9, is planned for launch in 2013-14. To be operational in 2014, GAGAN with its enhanced accuracy will not only make the skies and landings safer by providing three-dimensional approach operation but also enable less safety-spacing between two aircraft allowing for thrice more aircraft to fly. Better air space management is estimated to cut airline fuel costs by 20 per cent[96] providing for cheaper air travel. As GAGAN's footprint extends from Middle East Asia to Australia and GAGAN being interoperable with the GPS, the augmentation can be provided over this region too. GAGAN could bridge the gap between the navigation systems of Europe (served by EGNOS) and Japan (served by MSAS). Even though designed primarily for civil aviation applications, GAGAN is scalable enough to provide far-ranging enhanced end-user services like agriculture, land mapping, emergency response, natural resources, mining and vehicle tracking.[97]

In yet another giant leap towards self-reliance, the Indian version of the US' GPS, the Indian Regional Navigational Satellite System (IRNSS) is being fielded by ISRO. Approved in 2006, the project envisages a constellation of 7 navigation satellites, 3 in geostationary and 4 in geosynchronous orbits (inclined 29° with the equatorial plane), beaming navigation signals in the S band (2-4 GHz) and ranging signals in the C band, providing 10 m accuracy over the mainland.[98] The first satellite of the IRNSS constellation, the 1,425 kg IRNSS 1A, was launched onboard the PSLV C22 into a geosynchronous orbit in July 2013 while the full constellation is planned to be operational in 2014[99] but likely to roll over to 2015. The project includes complete indigenous content in the space and ground segments as well as in the end-user receivers. Self-reliance comes to the fore not only in fielding the entire gamut of indigenous infrastructure towards this service, but, more importantly, in cutting down reliance on navigation services from foreign providers in times of crises.

In its mid- and long-term vision, ISRO sees for itself performing a stellar role in two fields. The first aims at lowering the cost of orbiting a satellite and the second involves active participation in the space sciences that centre around missions to the Moon and Mars, including manned ones. Towards the former, the need arises from the fact that each satellite that is launched into a LEO by the PSLVs and GSLVs, costs US$ 6,000-7,000 per kg. To maintain commercial viability and global competitiveness, the Space Vision 2025 strategy enunciates the use of a Reusable Launch Vehicle (RLV) that is based on a Two-Stage-To-Orbit (TSTO) configuration.[100] The design envisages the first stage to attain a speed of Mach 12, twelve times the speed of sound, separate and fly back to the launch pad while the second stage continues onwards to orbit the satellite.[101] Called Avatar, the RLV would weigh 25 tons and eventually bring down the cost of orbiting a one ton satellite to US$ 67 per kg.[102]

For putting humans into space, ISRO has already achieved success with its 550 kg space Capsule Recovery Experiment 1 (SRE 1) launched aboard the PSLV, which orbited for 12 days and was safely recovered on the Earth in January 2007. This is obviously the first step towards realising the goal of an Indian in space, launched and recovered safely aboard indigenous spacecraft. The next step would well be a crew of two or three men in a fully autonomous orbital vehicle on a 300 km LEO.[103] As per an agreement with Russia in 2008, an Indian cosmonaut will receive training on a Russian Soyuz flight in 2013, followed by an Indian-only three-man crew aboard a capsule launched by the GSLV in 2015.[104]

In the realm of exploring space beyond the Earth, Chandrayaan 1's remarkable success has bolstered ISRO's confidence in planning for Chandrayaan 2 to the Moon in 2013. The Lander will be provided by Russia while the Orbiter and Lunar Rover will be built by ISRO.[105] Mangalyaan, a mission to Mars aims to place a 15 kg scientific payload on a 500 x 80,000 km orbit around the red planet.[106] Launched on November 5, 2013, Mangalyaan in the country's first inter-planetary mission, making the country the fourth in the world to do so.

The Strategic Focus

In keeping with Dr. Vikram Sarabhai's vision of Indian space technologies being applied purely for the common good of its people, the roots of the Indian space programme lay in peace-oriented 'economic and societal growth' and the remarkable success achieved therein enabled those very tools of development to propel the programme to grow beyond what was initially conceived, and bear fruit in providing the country a place on the global spacefaring platform. The space programme has matured into a multi-dimensional one where its operationalisation has benefited diverse stakeholders such as the government, users among the general public, industrial suppliers, customers, academia, foreign space agencies and international organisations. The programme today stands robust and self-reliant, contributing significantly to the national economy, an economy that is growing and stamping its presence at the global level.

At the global geostrategic level, it would be prudent for India to focus on two growth areas that will cement its stand in the international arena in the years to come. First and foremost, within the ambit of its policy of strategic autonomy should be the national realm – in the field of safeguarding the country's space interests. Ranking sixth in the world in terms of space budget and technological capabilities, with over US$ 26 billion invested in space assets,[107] India can ill afford to ignore the practical aspects of ensuring the security of its space infrastructure, especially of its space-based assets. With space infrastructure providing a significant chunk to the country's economy, a pragmatic threat assessment first needs to identify vulnerable areas in the space matrix, from which should flow mechanisms that need to be put in place to prevent an overt or covert attempt at breaching this architecture.

The second growth area should be in the field of international relations, where the need emphasises a larger and an all-encompassing Indian presence in the global field of partnerships and collaborations. While the country should continue with capability building and enhancements in the space sector for domestic needs, there is much credit in exploiting its growing economic might and further enlarging geopolitical gains and spinoffs by engaging with foreign partners. This could be realised by a two-pronged strategy. The first should aim at joint ventures with countries which have not yet "arrived" at the space arena, countries that have modest budgets but relatively larger space ambitions. Such countries should be proffered the entire range of services that Antrix, the marketing arm of ISRO, currently engages in, with a view to project and expand the country's "soft power". Significant inroads can still be made and widened in the field of providing imagery, assistance in constructing satellites and ground infrastructure, and, of course, the launch services. The second prong should aim at bilateral and multilateral partnerships with nations that have already established themselves as major players in space, with the aim of sharing emerging technologies and trends for mutual benefit and energising cooperative ventures in space sciences and in space exploration missions beyond the Earth. It would also serve the Indian interests well to remain actively engaged in the international fora on discussions and negotiations on space security and the international space law, with a view to remain abreast of the dynamic and ever-evolving power balance.

While the security imperatives that impinge upon space assets are being dealt with separately in this book, this chapter concludes with the thought that it would be in India's interests that its growth narrative includes a comprehensive space security policy in text and practice, from which should evolve key drivers that should chart the course for a national space security framework.

Notes

1. Dr. Man Mohan Singh, "Of Oxford, Economics, Empire and Freedom," *The Hindu*, July 10, 2005.
2. Gurcharan Das, *India Unbound* (Noida, India: Penguin Books, 2007), pp. x-xi.
3. Pikay Richardson, "New Science, Technology and Innovation Developments in India," April 23, 2002, accessed at ftp://ftp.cordis.europa.eu/pub/improving/docs/sstp_strata_workshop_session1_richardson.pdf on July 20, 2012.
4. "India GDP," http://www.tradingeconomics.com/india/gdp, accessed on July 20, 2012.
5. "India: Country Results Profile", accessed at http://web.worldbank.org/WBSITE/EXTERNAL/NEWS/0,,contentMDK: 22888405~menuPK:141310~pagePK:34370

~piPK:34424~theSitePK:4607,00.html on July 16, 2012.

6. Accessed at http://povertydata.worldbank.org/poverty/country/IND on July 23, 2012.
7. Samuel Kornstein and Paul Artiuch, "Battling Food Waste in India," January 14, 2012 accessed at http://inpec.in/2012/01/14/battling-food-waste-in-india/ on July 23, 2012.
8. Satyajit Das, "The Great Pretender: India's Economic Past and Future: Part 1, 'India Shining'," accessed at http://www.economonitor.com/blog/2012/05/the-great-pretender-indias-economic-past-and-future-part-1-india-shining/ on July 23, 2012.
9. Subhash Kak, "History of Indian Science," July 31, 2002, accessed at http://www.ece.lsu.edu/kak/grolier.pdf on July 30, 2012.
10. Richard Thompson, "Planetary Diameters in the Surya-Siddhant," *Journal of Scientific Exploration,* vol. 11, no. 2, 1997, p.196.
11. "Indian Astronomy," accessed at http://www.spacetoday.org/India/IndianAstronomy.html on July 30, 2012.
12. "Aryabhatta Biography," accessed at http://www.biography.co.in/aryabhatta-biography.html on July 30, 2012.
13. "Classical India," accessed at http://www.britannica.com/EBchecked/topic/179408/education/47452/The-introduction-of-Buddhist-influences#toc47453 on July 30, 2012.
14. "Aryabhatta Biography," accessed at http://www.biography.co.in/aryabhatta-biography.html on July 30, 2012.
15. "Indian Astronomy," accessed at http://www.spacetoday.org/India/IndianAstronomy.html on July 30, 2012.
16. "History of the Physical Sciences in India" accessed at http://india_resource.tripod.com/physics.htm on July 31, 2012.
17. R.K. Kochhar, "Science in British India – I: Colonial Tool," *Current Science*, vol. 63, no. 11, December 10, 1992, p. 670.
18. R.K. Kochhar, "Science in India – II: Indian Response," accessed at rajeshkochhar.com/data/publications/ ScienceinBritishIndia-II.pdf on August 1, 2012.
19. David Arnold, *Science, Technology, and Medicine in Colonial India* (Cambridge: Cambridge University Press, 2000).p. 169.
20. Roddam Narsimha, "Science, Technology and the Economy: An Indian Perspective, " accessed at http://repository.ias.ac.in/24528/1/324.pdf on August 10, 2012.
21. http://www.dst.gov.in/about_us/intro_DST.htm, accessed on August 14, 2012.
22. Richardson, n. 3.
23. "CSIR Annual Report 2009-10," accessed at http://www.csir.res.in/External/Utilities/Frames/aboutcsir/main _page.asp?a=topframe.htm&b=leftcon.htm&c=../../../Heads/aboutcsir/about_us.htm on August 30, 2012.

24. Narsimha, n. 20.
25. Percival Spear, *India: A Modern History* (Ann Arbor, Michigan: University of Michigan, 1961), p.444.
26. Rajeev Lochan, "Some Reflections on Collective Security in Space," in John M. Logsdon and James Moltz, eds., *Collective Security in Space: Asian Perspectives* (Washington, D.C.: George Washington University, Space Policy Institute, 2008), p.33.
27. Dinshaw Mistry, "The Geostrategic Implications of India's Space Program," *Asian Survey 41*, no. 6, November/December 2001, p. 1042.
28. A modest budget for space is a relative term as it is 7.5 percent of NASA's budget, a quarter of the European Space Agency's and a mere 0.34 per cent of the Indian government's expenditure. "Winning is a Matter of Belief: ISRO Chief," *The Hindu*, September 9, 2012, accessed at http://www.thehindu.com/todays-paper/tp-features/tp-downtown/winning-is-a-matter-of-belief-isro-chief/article3875930.ece, on September 26, 2012.
29. Rs 6,715 crore has been allotted to ISRO for the year 2012-13. "Still Eyeing Space: Higher Budget Support for ISRO is Welcome," *Business Standard,* April 3, 2012, accessed at http://www.business-standard.com/india/news/still-eyeing-space/469871/ on September 25, 2012.
30. Bharath Gopalalswamy, "Indian Space Policy: Aiming Higher," *Space News*, July 21, 2008, p. 23.
31. "Tricolour's 4th National Flag on Moon," *The Economic Times,* November 15, 2008.
32. Mark Williams, "India's Space Ambitions Soar," *MIT Technology Review*, July 30, 2007, accessed at http://www. earth2orbit.com/pdf/ISRO.PDF on September 17, 2012.
33. James Clay Moltz, *Asia's Space Race: National Motivations, Regional Rivalries And International Risks* (New York: Columbia University Press, 2012), p. 112.
34. B. N. Suresh, "History of Indian Launchers," *Acta Astronautica 63,* 2008, p. 428.
35. Brian Harvey, Henk H. F. Smid and Theo Pirard, *Emerging Space Powers: The New Space Programs of Asia, the Middle East, and South America* (Chichester, UK: Praxis Publishing Ltd., 2010), p. 144.
36. Ibid., p. 142.
37. Moltz, n. 33, p. 114.
38. Suresh, n. 34, p. 429.
39. Harvey, et. al., n. 35, p. 146.
40. In a span of five years, the rocket launch weight went from 60 kg to over two tonnes. http://www.bhaskarastro. org/earlyhistory.htm, accessed on October 1, 2012.
41. "First Indian Rockets," accessed at http://www.bhaskarastro.org/rocketsfirst.htm on October 1, 2012.

42. Williams, n. 32.
43. V.S. Mani, "Space Policy and Law in India and Its Relevance to the Pacific Rim," *Journal of Space Law* 35, no. 2, Winter 2009, pp. 618-619.
44. "History of Indian Space Programme," accessed at http://www.bhaskarastro.org/earlyhistory.htm on October 3, 2012.
45. Harvey, et al., n. 35, p. 149.
46. The solid propellant was made at SHAR and the Americans chose similar composition decades later for their boosters in the Space Shuttle.
47. Harvey, et al., n. 35, pp. 164-169.
48. ISRO, "Launch Vehicles," accessed at http://www.isro.org/Launchvehicles/launchvehicles.aspx on October 5, 2012.
49. Harvey, et al., n. 35, pp. 202-204.
50. "India's Spy Satellite Boost", *BBC News*, November 27, 2001, accessed at http://news.bbc.co.uk/2/ hi/south_asia/1679321.stm on October 15, 2012.
51. The GSLV became the first and also the only geosynchronous rocket in the world to have a solid-fuelled first stage. Harvey, et al., n. 35, p. 221.
52. The liquid-propelled Vikas engine is the Indian *avatar* of the French Viking engine, whose technology was acquired in the 1970s. P. V. Manoranjan Rao and P. Radhakrishnan, *A Brief History of Rocketry in ISRO* (Hyderabad: Universities Press (India) Private Limited, 2012), pp. 144-145.
53. A cryogenic rocket engine, the most complex of all propulsion systems, is an engine that uses cryogenic fuel or oxidiser, that is, its fuel or oxidiser (or both) are gases stored at cryogenic temperatures (below -150 degrees Celsius) in liquid form, as against using gases in pressurised form. Liquid hydrogen (LH2) and liquid oxygen (LOX) is the most commonly used composition. The complexity lies in fuel management due to their low boiling points. India is the sixth country in the world to have used them.
54. India had tested its new Agni ballistic missile in 1989, a missile that centred around the first stage rocket of the SLV-3.
55. Prof. U.R. Rao, former ISRO Chairman, *Space India,* October 1993-March 1994, accessed at http://www.frontlineonnet. com/fl2710/stories/20100521271010100.htm on October 8, 2012.
56. Harvey, et al., n. 35, pp. 223-225.
57. The financial aid package was essentially US$ 400 million that the USA paid to Russia for seven American flights to the Mir space station, the very exact amount that the Russians lost in the Indian cryogenic deal. Harvey, et al., n. 35, p. 224.
58. Group Captain Joseph Noronha, "India's Space Endeavour, " *Indian Defence Review*, vol. 26, no. 3. July/September 2011, p. 22.
59. ISRO, "GSLV Mark III," accessed at http://www.isro.org/Launchvehicles/

GSLVMARKIII/mark3.aspx on October 8, 2012.

60. Rao and Radhakrishnan, n. 52, p. 318.
61. A total of 1,200 hours of such programmes made by the Space Applications Centre, Ahmedabad, was beamed up from the country's first ground station at Ahmedabad, to ATS-6 and retransmitted back for four hours every day.
62. Harvey, et al., n. 35, pp. 148-157.
63. "History of Indian Space Programme," accessed at http://www.bhaskarastro.org/earlyhistory.htm#a10 on October 10, 2012.
64. Harvey, et al., n. 35, pp. 184-185.
65. n. 65.
66. Harvey, et al., n. 35, pp. 159-164.
67. ISRO, "Earth Observation Satellites," accessed at http://www.isro.org/satellites/irs-1a.aspx on October 15, 2012.
68. Harvey, et al., n. 35, pp. 174-175.
69. Ibid., p 175.
70. "India's Spy Satellite Boost", *BBC News*, November 27, 2001, accessed at http://news.bbc.co.uk/2/ hi/south_asia/1679321.stm on October 15, 2012.
71. ISRO, "Cartosat 2B," accessed at http://www.isro.org/satellites/cartosat-2b.aspx on October 15, 2012.
72. "Why RISAT 2 Came Before RISAT 1," May 3, 2012, accessed at http://www.strategypage.com/htmw/ htspace/20120503.aspx on October 18, 2012.
73. n. 67.
74. n. 63.
75. Harvey, et al., n. 35, pp. 184-189.
76. ISRO, "INSAT 1B", accessed at http://www.isro.org/satellites/insat-1b.aspx on October 16, 2012.
77. Harvey, et al., n. 35, pp. 190-195.
78. Ibid., p. 196.
79. ISRO, "Telemedicine," accessed at http://www.isro.org/scripts/telemedicine.aspx on October 17, 2012.
80. Harvey, et al., n. 35, p. 220.
81. ISRO, "All Satellites," accessed at http://www.isro.org/satellites/allsatellites.aspx on October 17, 2012.
82. Committee on the Peaceful Uses of Outer Space, "Actual Situation in the Geostationary Orbit," 49th Session, Vienna, February 06-17, 2012, accessed at http://www.unoosa.org/pdf/limited /c1/AC105_C1_2012_CRP25E.pdf on October 17, 2012.
83. Launches from SHAR are directed at 135 degree azimuth, that is, towards the south-east. PSLVs after launch are steered around Sri Lanka and settled into

their polar orbits. GSLVs ideally need to be launched as eastwards as possible, that is, as closer inclined along the Equator as possible. Launching the GSLV at 135 degree and then redirecting the geostationary satellite to inch towards the equatorial orbit, carries with it the associated penalty of increasing the propellant required to be stored in the satellite, that comprises on the useful payload that a satellite can carry. Given this constraint, the GSLV Mk II can deliver at best a 2,500 kg satellite into GEO. The GSLV Mk III was conceived with the aim to address just this issue, that of launching satellites of up to five ton class. Rao and Radhakrishnan, n. 52, pp. 180-190.

84. ISRO, "Tracking Facility"" accessed at http://www.isro.org/GroundFacilities/trackingfacility.aspx on October 18, 2012.
85. Press Information Bureau, Government of India, "Setting up of Indian Institute of Space Science & Technology," April 26, 2012, accessed at http://pib.nic.in/newsite/erelease.aspx?relid=27199 o0n October 18, 2012.
86. Moltz, n. 33, p. 121.
87. Harvey, et al., n. 35, p. 250.
88. "Overview of Indian Space Sector 2010," accessed at http://mycii.in/KmResourceApplication/E000000056. 662.Antrix-CII-Deloitte%20report%20on%20Indian%20Space%20Sector%202010.25Aug10.pdf on October 19, 2012.
89. Antrix, "Company Overview," accessed at http://www.antrix.gov.in/aboutus.html on October 19, 2012.
90. Antrix, "Annual Report 2009-2010," accessed at http://www.antrix.gov.in/Antrix_annual_report_09_10.pdf on October 19, 2012.
91. Astrium is an aerospace subsidiary of the European Aeronautic Defence and Space Company (EADS) that provides civil and military space systems and services.
92. n. 88.
93. "Indian Space Agency Wants Second Rocket Assembly Facility," accessed at http://www.space-travel.com/reports/ Indian_space_agency_wants_second_rocket_assembly_facility_999.html on July 10, 2013.
94. "ISRO to Launch 58 Space Missions by 2017," October 8, 2012, accessed at http://www.satellitetoday.com/st/ headlines/39643.html?hq_e=el&hq_m=2537808&hq_l=22&hq_v=7c49e30760 on October 25, 2012.
95. ISRO, "Future Programme," accessed at http://www.isro.org/scripts/futureprogramme.aspx on October 26, 2012.
96. "Air Navigation System GAGAN to Help Jet Airways, Indian Airlines, Others to cut Fuel Cost by upto 20%," *The Economic Times*, October 8, 2012, accessed at http://articles.economictimes.indiatimes.com/2012-10-08/news/34323085_1_indian-skies-air-navigation-airlines-end on October 26, 2012.
97. A.S. Ganeshan, ISRO, "GAGAN is Expected to Replace the GPS Receivers and

Provide Data Integrity," February 2012, accessed at http://mycoordinates.org/gagan-is-expected-to-replace-the-gps-receivers-and-provide-data-integrity/ on October 26,2012.

98. A. Bhaskaranarayana, ISRO, "Indian IRNSS and GAGAN," presentation to COSPAR meeting at Montreal, July 15, 2008, accessed at http://www.oosa.unvienna.org/pdf/icg/2008/expert/2-3.pdf on October 29, 2012.
99. ISRO, "Future Programmes," accessed at http://www.isro.org/scripts/futureprogramme.aspx on October 29, 2012.
100. ISRO, "Space Vision India 2025," accessed at http://www.isro.org/vision.aspx on October 29, 2012.
101. Rao and Radhakrishnan, n. 52, pp. 318-319.
102. Susmita Mohanty, "Indian Space Programme," June 2008, accessed at http://www.earth2orbit.com/ pdf/ISRO.PDF on October 29, 2012.
103. n. 99.
104. K. S. Jayaraman, "India Plans First Manned mission with Assistance from Russian space Agency," *Space News*, February 2, 2009, p. 11.
105. n. 99.
106. Press Trust of India, "Manmohan Formally Announces India's Mars Mission," *The Hindu*, August 15, 2012, accessed at http://www.thehindu.com/sci-tech/science/article3775271.ece on October 31, 2012.
107. Arvind Gupta, Amitav Mallik and Ajey Lele, eds., *Space Security: Need for Global Convergence* (New Delhi: Pentagon Press, 2012), p. vii.

5

Space Assets: Threats and Vulnerabilities

Our scientific power has outrun our spiritual power. We have guided missiles and misguided men.

— Martin Luther King, Jr.

February 10, 2009, marked the day when two satellites in orbit, an American satellite, the Iridium 33 and a Russian satellite, the defunct Kosmos 2251, accidently collided 790 km above Siberia. While eight such collisions between orbiting platforms had been recorded prior to this accident, this was by far the biggest incident of its kind till then.[1] Though a threat to satellites from another orbiting platform is considered minimal, it brings to the fore that such a threat does exist and will only increase exponentially as more and more countries launch satellites in the years to come. The threat to a satellite that emanates from other satellites in similar orbits gets further compounded by the space junk or debris that such a collision creates. There are thousands of pieces of orbiting debris that have the potential to further pose a collision risk to other operational satellites. While this debris was created by an accident, there have been instances of intentional operations conducted in space that have led to the creation of unwanted debris. Anti-Satellite (ASAT) weapons, designed to destroy satellites, were tested both by the USSR and USA in the period 1960s to 1980s, creating debris as a result. While the two superpowers thereafter applied a self-imposed moratorium on further testing of ASAT systems, China's ASAT test in 2007 added another 35,000 pieces of one centimetre size or larger to the ever growing space junk,[2] considering that even a centimetre sized debris travelling at orbital speeds into a satellite is sufficient to render it inoperable.

This chapter offers an insight into threats that are posed to orbiting satellites, both from the naturally existing cosmic environment and by man-made causes. Each threat will also be analysed briefly from the perspective of how vulnerable an individual space platform is from the particular threat, and the measures and techniques adopted towards safeguarding the space assets.

Threats from the Cosmos

Space Radiation

Unlike the common perception of space being a vast and frigid emptiness, space as is now known today, is a vacuum though with an extremely low density of particles but with considerable cosmic activity. Due to the extremely low density of particles, space effectively offers no resistance to its movement of cosmic rays and electromagnetic radiation. Closer home, in the altitudes where satellites orbit, space is populated by electrically charged particles at very low densities, the motions of which are influenced by the Earth's magnetic field. The naturally existing environment is the Earth's trapped radiation belts, the Van Allen belts, solar particle events and galactic cosmic rays. These are made up of subatomic particles accelerated to intense energy levels by the violent activities on the Sun, as well as exploding stars throughout the Galaxy. These radiations comprise IR, UV, X-ray and gamma radiation, and in addition various energised charged particles. Different components of space radiation have different effects on spacecraft. Solar particles impacting upon a spacecraft's airframe could affect its motion and the continuous impact of solar particles can cause physical damage, especially to delicate components like sensors and solar panels. Individual charged particles can penetrate electronic components and could corrupt computer memory, while charged particles in sufficient numbers may induce noticeable currents or significant voltages, which on discharge could cause severe damage. Solar flares can damage the sensitive electronics of the satellite or trigger phantom commands in the onboard computers. In respect of manned flights, these rays are extremely hazardous to living cells, and space travellers exposed to them could suffer anything from near immediate acute radiation sickness to cancers in later life.

To mitigate the harmful effects of these radiations, different shielding materials like tin, steel, copper, lead and boron carbide applied on vulnerable components of a spacecraft provide varying protective effectiveness against different kinds of radiation, while aluminium is very commonly used for the walls of the spacecraft. Manned spacecraft, of course, come with much more stringent protocols in respect of foolproof shielding while the space suits' outer protective layer and the visor on the helmet provide shielding against radiation to enable space travellers to conduct limited-duration out-of-vehicle operations.

Cosmic Bodies

Cosmic and solar activity impacting the Earth is not restricted to the radiations discussed above. The Earth is frequently bombarded with solid cosmic matter from the inter-galactic space and from our own solar system. Comets are large chunks of ice, rock and dust that originate from deep space, and on entering the solar system, have their frozen mass vaporised due to the Sun's radiation, leading to formation of the glowing comet tail. A few of them are attracted by the gravitational pull of the bigger planets of the solar system and enter the solar system orbiting around the Sun. While the tail may extend to millions of kilometres, the diameter of the core body of the comet itself is a few tens of kilometres.

Asteroids, the subject of numerous science fiction films, on the other hand, are constituents of our solar system itself. They are solid, rocky, tailless bodies that orbit the Sun in elliptical orbits taking three to six years to complete the circuit around the Sun. They vary in size from between 940 km diameter to pebble sized ones that measure a few centimetres. Some of the asteroids and comets that have their orbits coming in close proximity to that of the Earth's are classified as Near Earth Objects (NEO). Their orbits are followed by Earth-based remote sensing, by both optical and radar tracking systems, as an asteroid of one km diameter or more, impacting the Earth could cause extinction of the life-form from this planet. A total of 911 "killer asteroids" have been traced as of 2011, while 70 are estimated to be yet-to-be-discovered.[3] It would be futile to discuss the threat that such asteroids and comets pose to satellites, as mankind is likely to be consumed by issues of far more cataclysmic proportions at the global level. At present, no known NEO asteroid or comet has an orbit that could impact the Earth in the first half of this century.[4] There are, however, estimated to be a million asteroids that are 100 m or less in size, which on impact with a satellite, would totally obliterate it.

In addition, the planet receives a fair share of meteoroids, small rocky pieces that broke off from asteroids and comets at some point in time. They range in diameter from micrometres to kilometres. The smaller ones, on entering the Earth's atmosphere are, heated to incandescence and produce the meteor glow. The larger meteoroids normally survive entry into the atmosphere and the unvaporised chunks, on impact with the Earth's surface, are called meteorites. It is estimated that about 10,000 tons of these cosmic rocks fall on the Earth annually. Notwithstanding this alarming statistic, small meteoroids and micrometeoroids large enough to damage spacecraft are considered "somewhat rare" by space scientists. For example, 25,000 years is the estimated time between collisions between a one gram meteoroid and a satellite in LEO.[5]

Before analysing the mitigation techniques that are employed to overcome or avoid collision threats that such cosmic bodies pose to satellites, it would be essential to first look at systems that detect and track such threats. Most major players in space are at various stages of their Spaceguard programmes, a programme that aims at discovering NEOs and tracking their orbits with respect to that of the Earth. Ground-based telescopes and radar tracking systems support the Spaceguard survey while similar functions are also being carried out by NEO-tracking satellites. The response against a detected threat varies according to the warning time available. In the case of an asteroid or comet that has the potential to cause global catastrophe, the mitigation technique would aim to either fragment and destroy the incoming cosmic body or deflect the threat by changing the velocity of the body. In the case of much smaller bodies that could have little or insignificant impact on the Earth's surface, but are significant enough to destroy a satellite in its collision path, however remote the chances, it would be more prudent for the satellite to be manoeuvred and steered away from the collision path by firing its onboard jets or thrusters. These thrusters are actually employed for station-keeping, that is, for keeping the satellite on course in its orbit station, and consuming a part of the critical and limited fuel for disaster mitigation would tend to reduce the satellite's operational life, which, of course, would be a very small price to pay for keeping the satellite alive.

To safeguard spacecraft from even smaller bodies — those micrometeorites that do not have the potential to destroy a satellite but a stream of which could affect the structure — a Whipple shield is employed to provide them the protection. This foil film, normally of aluminium, is placed a short distance in front of the spacecraft to absorb the initial impact of the micrometeorite collision. The full force of the impact causes the micrometeoroid to vaporise into plasma that quickly spreads across the film's surface. By the time the plasma crosses the gap between the shield and the spacecraft's structure, it is too diffused to cause any significant damage. Ceramic fabric woven shields and multi-layer flexible fabric are relatively newer trends in the Whipple shield which offer increased protection with reduced weight, and offer some protection against space debris too.

Threats Emanating from Man-made Activities

Cerise, a 50 kg French military reconnaissance satellite in LEO, was hit and damaged in 1996, by a catalogued space object piece of debris, a fragment from, ironically, a French Ariane rocket that had exploded in space ten years earlier. This was the first verified collision between an operational satellite and space debris.[6]

Almost six decades of space travel and exploration by an increasing number of countries have seen an ever-increasing deployment of space assets in orbit. These assets may have been deployed to serve the country's interests in the ever burgeoning fields of space-based services, applications and national security, but their deployment has also in a large part contributed incidentally to threats to these very space assets. As if hazards posed from the natural debris, that is, the cosmic bodies were not enough, space, in the region where satellites orbit, is littered with space junk that comprises spent rocket stages, non-operational satellites and broken parts resulting from collisions, amongst others. While such orbiting objects have been a technological byproduct of the processes of launch and orbit of satellites, there are some nations that have invested in testing and deployment of weapon platforms with the intentional aim of destroying or degrading other satellites in times of need. And the problem is only getting worse.

Space Debris

Space debris or space junk is a generic term that encompasses all forms of derelict man-made objects in space which have the potential of becoming a hazard to space travellers and spacecraft. Also called orbital debris, as, unlike a natural meteoroid which is a one-time transient phenomenon passing through the atmosphere, man-made debris remains in orbit for its life-time. These consist of spent upper-stage rocket boosters, dead satellites, rogue satellites, paint flakes, coolant released from nuclear powered satellites, and bits and parts arising out of erosion, corrosion and collision fragments over a period of time, all travelling at orbital speeds crossing each other at relative speeds of 10 km per second, in varying orbits at various altitudes. The list of items that constitute space debris, surprisingly, also includes items lost by astronauts while conducting their activities in space. Items recorded as lost and orbiting in LEO include two cameras, a glove, a tool bag, a pair of pliers, a wrench and garbage bags jettisoned from the space station Mir. More damaging are the fragments that result from collisions between a satellite and a piece of debris, or worse, between two satellites. The collision between the USA's 560 kg Iridium 33 satellite and the Soviets' defunct 950 kg Kosmos 2251 satellite in February 2009, occurring at a relative speed of impact of 11.7 km per second, is estimated to have created 10,000 fragments varying in size from centimetres to tennis-ball size,[7] that makes up about 10 per cent of all known space debris accumulated in the last 55 years.[8] Another contributor to the ever increasing numbers of orbital flotsam is the debris caused by ASAT testing. Commencing in the 1950s, both the USA and USSR conducted ASAT tests through various platforms — ground based, air-launched as well as co-orbital satellite-based. These tests continued right through the 1980s yielding various success rates, destroying in the process several of their targets which were dead or decaying satellites. It was at this stage that, realising the hazard of space debris on orbiting satellites, both the superpowers imposed on themselves a moratorium on further ASAT testing.

This state of affairs continued till 2007 and the status changed when China launched a ground-based missile interceptor at its weather satellite, the 750 kg FY-1C, destroying the same at an altitude of 865 km. NASA, as of June 2012, estimated the test having created a debris cloud containing 150,000 pieces larger than one centimetre, moving outwards from the point of impact in polar orbits at altitudes between 200 and 4,000 km, of which 79 per cent of the pieces will remain in orbit for 100 years.[9] The Chinese ASAT test was followed a year later by the US firing a ground-based interceptor to destroy a defective spy satellite, the USA-163, orbiting at 250 km. Due to the low altitude of destruction where the atmospheric drag is relatively higher, estimates indicate that the debris decayed in orbit and none remained 40 days after the event.[10]

Size Does Matter: A vast majority of the estimated tens of millions of pieces of space junk are smaller particles, less than one centimetre in size. Such pieces, on collision, do not pose the risk of causing significant structural damage to space assets but cause erosion and corrosion to the spacecraft's structure. A large part of such a hazard is obviated by the Whipple shield, as discussed in the preceding section. Debris that is of concern is in the range of one centimetre and above, more than 300,000 pieces of which orbit in the LEO[11] which have the potential to destroy satellites on collision.[12]

Kessler Syndrome: Space is large and voluminous. The entirety of LEO (up to 2,000 km altitude) has a volume so large that for, say, 1,000 operational satellites, there would be room of 600 million cubic km for each, enough for spacecraft to freely move about their orbits without conflicting with any other spacecraft and more than adequate enough to launch and orbit more satellites. While the likelihood of accidental collision between two satellites is rare, it is the threat posed from space debris that caps mankind's ambitions in space. Debris of various sizes orbiting at various altitudes in LEO (up to 2,000 km) would total up to a mass of 3,000 tons.[13] The figure by itself is not alarming as the threat is not from a single unit of that mass but from the millions of pieces that make up that mass. Each collision between a piece of debris and a spacecraft would cause thousands of pieces of new debris, each of which then has the potential of posing a threat and increasing the likelihood of collision with other satellites. This cascading effect, called the Kessler effect, further adds to litter in space raising the hazard-bar exponentially. A case in point is the reported hit to a Russian satellite, the BLITS, on January 22, 2013, by a piece of debris that resulted from the Chinese ASAT test of 2007.[14]

Debris and Orbit Bands: The presence of debris and the hazards it poses is different in different orbit bands. LEO remains the most cluttered, with both operational satellites and space debris. This is also the altitude that space travellers visit, including almost permanent manning of the International Space Station (ISS). The lower end of LEO, up to 400 km, is a relatively safer region as debris experiences a certain amount of atmospheric drag, decays faster, burning through the atmosphere and landing back on the Earth's surface at an average of once a day. With a large number of satellites in Sun-synchronous orbits at varying inclinations that carry out 15 orbits a day typical of high-speed LEO missions that provide global coverage, approaches near the debris slots are far more frequent. The polar regions are common crossing points for polar orbits and, thus, have a larger density of satellites. NASA estimates that there is ten times more debris per cubic km at latitudes greater than 75 degrees than there is above the Equator.[15] It is the LEO region where the Kessler effect would be most likely and pronounced. Within the LEO, the altitude band between 800 to 1,500 km is the most dense, with debris with very large orbit life. Both China's ASAT test of 2007 and the collision of the US and Soviet satellites in 2009, took place in this altitude band, contributing hugely to the debris already present. The Geostationary Earth Orbit (GEO) at 36,000 km has its own peculiarities in respect of debris. Satellites in GEO share the same orbital path, clustered together over their 'piece of ground'. Dead satellites become a concern as they can no longer be manoeuvred out of harm's way and in addition become a collision risk themselves. On the upside, since the orbital speeds are relatively slower as compared to satellites in LEO, collision with a piece of debris would still knock it out of operation but the debris produced would be significantly less.

Debris Tracking: A reliable and accurate catalogue of space debris is a fundamental requirement towards avoiding and mitigating the hazards that arise from space debris and involves identifying and tracking the individual pieces of debris. The data on such hazards is collected, from both ground-based radar, laser and optical telescopes and space-based telescopes and contributes to the Space Situational Awareness (SSA). Current technology enables tracking of objects 10 cm or larger while the need is also to catalogue those that are of 1-10 cm size, a majority of which remain unobserved. As the object is tracked, its trajectory is predicted and calculated to arrive at its 'point of impact' with an orbiting satellite. To alleviate the threat posed by space debris, the US Strategic Command's Joint Space Operations Centre (JSPOC) detects, tracks, and identifies space objects through an elaborate network of 29 ground-based radars and optical sensors deployed around the world. In addition to protecting US spacecraft, JSPOC extends this capability—at no cost to the international community—by warning countries and commercial space operators when their satellites are at risk from large space debris or other satellites.[16]

Mitigation Techniques: Whipple shields on satellites provide protection from small sized debris of 5-10 mm. Larger pieces of debris, on being tracked and predicted to be on a collision course with a satellite, would necessitate orbital manoeuvring to steer the satellite away from its trajectory and bring it back on course once the threat flies past. This, of course, would be the most direct procedure to avoid a collision. Most of the spacefaring nations are increasingly becoming aware of the mitigation techniques that aim to avoid such an accident in the first place. 'Litter-free' considerations are incorporated by most, in the design stage itself where new spacecraft are designed with provisions for retrieval or removal at the end of their operational life. While in GEO, satellites are mandated to be kept within the boundaries of their 'station keeping box' which is defined by the International Telegraph Union (ITU) to be within ± 0.1 percent of the slot in longitude and latitude, which for 36,000 km translates to a swath of 148 km for a satellite in the East-West direction – an arc large enough to provide safety against collision and prevent interference in communications.[17] Spent rockets and satellites in LEO nearing the end of their lives that have sufficient propellant left unconsumed, can be powered back onto the Earth in a controlled reentry. In the case of insufficient propellant, these satellites could be powered into decaying orbits, that is, orbits in lower altitudes where atmospheric drag then deorbits them after some time. GEO satellites at the end of their use could be powered onto a higher 'graveyard orbit', 235 km higher than the GEO, making them safe as debris themselves and in addition, decluttering the GEO space. Global space players are actively engaged in tackling the menace that space debris poses and will continue to pose in the years to come. While current efforts have focussed on reducing the creation of future space junk, the space community is seized of the necessity of removing large pieces of existing space debris. Spearheaded by NASA, several proposals are in consideration for debris-management and removal. Amongst ideas that have yet to reach cost-effective fruition is one involving robotic spacecraft to capture debris and bring it back to the Earth, or refuelling dead satellites to power them into graveyard orbits. NASA's Orbital Debris Programme Office has also toyed with the idea of using ground-based laser on the debris, aiming to slow down the junk and reduce its orbital altitude to that where atmospheric drag brings the junk down. Deploying solar sails and tethers at the end of a satellite's life to slow it down and burn it up on reentry are also some ideas of technologies that are yet to mature and which may see the light of the day in the years to come.

Hard-Kill Operations

Space is no longer a sanctuary.

- Michael Wynne

Secretary of the US Air Force, 2007

Militaries of spacefaring nations use space as a force multiplier, a capability that enhances tremendously the efficiency of other military assets, in pursuance of the country's military objectives. Military satellites and dual-use satellites today are employed in a vast spectrum of military activities that range from gathering intelligence, navigation, military communications and weather, to more aggressive ventures like guiding precision weapons and enhancing missile defence, and to space-denial activities of targeting satellites as part of counter-space operations. The threat to victim satellites is not only in its destruction, the hard-kill option in the form of ASATs, but also takes the form of soft-kill options that aim not to destroy but disable or degrade the operation of the satellite. The threat to satellite-based systems actually transcends the threat to satellites and includes that to ground support infrastructure as well.

ASATs: These could take the form of Kinetic Energy (KE) weapons that aim to incapacitate or destroy a satellite by impacting it with a high speed mass, or in the form of orbiting satellites themselves being used as ASATs. These KE munitions could be launched from the ground or be wing-launched from a high altitude aircraft, or could even be launched from an orbiting platform. China's ASAT test in 2007 was of the former category – a ground-launched missile that was guided onto the satellite by ground-based radars. Threats from ground-launched weapons come not just from countries that possess space-launch capabilities but also from countries that are armed with ballistic missiles. The second category is of satellites that are designed as ASATs and pre-deployed in space disguised as peaceful satellites, awaiting instructions to track and attack their prey. Also called co-orbital satellites, they are positioned into a staging orbit where they could occupy sleeper slots for days or even years, till activation. While such a capability can be attributed to a mere handful of countries, the threat envelope increases manifold on the inclusion of over 40 countries that possess the capability of operating micro-satellites. Micro- and nano-satellites are ideally suited to play the role of killer satellites, as they can be launched quickly from small mobile launchers, or covertly as secondary payloads aboard large boosters committed to some peaceful launch mission, or equally covertly, attached to a 'peaceful' mother-satellite. Once in orbit, they are difficult to detect and track due to their small size. Use of nuclear weapons in space against satellites forms the other end of the spectrum of ASAT operations. The resultant electromagnetic pulse from a nuclear explosion in space damages the electronics aboard the satellites with equally damaging effects on equipment on the Earth below it. The resultant radiation has the potential to severely damage satellites in the line of sight, and the lingering radiation effects on satellites not in the line of sight reduce their operational life in LEO from years to months. Such an option comes with a tremendous prohibiting penalty, that of its indiscriminate effects, in that, such operations do not differentiate between friendly and hostile satellites. The option to use a nuclear device in space, however implausible, would have to be within the umbrella of an all-out war involving a nuclear exchange between the warring countries and not aimed specifically against assets in space.

Directed Energy Weapons (DEWs): Also called beam weapons, DEWs are also classified as ASATs and do not rely on direct-impact through a projectile but operate from 'stand-off' distances. Technologically more advanced and operating at or near the speed of light, these weapons kill in a few seconds and can destroy multiple targets with no intelligence indicators associated with their employment. Laser weapons, the choice of science fiction, operating at or near optical wavelengths, project a high energy beam onto the surface of the target. While low power laser affects the electronics aboard the satellite, high power laser irradiates the target with persistent energy, causing severe degradation and destruction by heating, melting or vaporising the structure of the satellite. China's 'blinding' of American spy satellites overflying China in 2006 by ground-based lasers was an example of demonstrating a capability where it did not aim to destroy American satellites but to make them useless over Chinese territory.[18] Not only do these weapons come in the ground- and air-launched configurations, space-based lasers have been under consideration for a number of years. Radio-Frequency (RF) weapons include both ground- and space-based RF emitters and have a similar aim of disabling electronic components and sub-systems of a satellite by firing an intense burst of radio energy. Space-based particle beam weapons are another class of weapons that aim to disable electronics aboard a satellite. In this case, the weapon fires a beam of elementary or atomic particles that causes permanent disablement by radiating enough energy to overload the target's electronic systems. While these 'game changing' plethora of weapons remain in the realm of yet-to-mature cost-effective technologies, there are a few formal projects on such weapons, with a majority of funding on research rather than on supported or fielded systems.[19]

Soft-Kill Operations

Though well within the ambit of counter-space operations, soft-kill operations do not aim at destruction of satellites but in the realm of non-lethal weaponry, aim to render them inoperable for a certain amount of time. Such weapons employ electronic attacks that use electromagnetic energy to deny or degrade the use of the electromagnetic spectrum by the adversary's satellites. One component of such an operation is jamming which could target the command and communications links, both uplink and downlink, and also jam the onboard sensors of the satellite. Jamming is effective up to the time the satellite remains in the line of sight. Jamming is specifically effective against satellites in semi-synchronous and GEO orbits, satellites that define military applications like GPS and military communications operating largely in the critical UHF band. Politically motivated intentional jamming by state and non-state actors has also been reported against communications satellites.

Spoofing is a short technological step beyond jamming and aims to mislead the target by capturing, altering and retransmitting a communication signal to the recipient. The 'spoofer' then enters the target's network as an authorised user and can generate false commands to cause the satellite to malfunction or fail its mission. Spoofing is one of the most discreet and deniable forms of attacking a satellite system.[20]

Satellite communications and control systems in addition, are vulnerable to cyber attacks and hacking like any terrestrial computer-controlled system. Cyber attacks aim to either degrade the adversary's information or to collect intelligence about them. A successful hack enables the hacker to disrupt the services of the satellite, take over its control and even destroy the satellite. China for one has made rapid progress in this field, amply demonstrated by the report in November 2011 that China had 'full access' to NASA's computers that controlled as many as 23 satellites.[21]

Mitigation Techniques Against Counter-Space Measures

Hostile actions against a country's space-based infrastructure can at times be confused with natural phenomena of solar activity or with other events like collision with space debris or hardware or software failure in the systems, which may tend to hide hostile or malicious acts. Thus, any technique to reduce the vulnerability of space platforms would need to commence with space situational awareness and a realistic threat assessment.

Space Situational Awareness (SSA): The term SSA refers to the capability to view, follow and predict the physical location of natural and man-made objects in orbit around the Earth, with the objective of avoiding collisions. It centres on near-time surveillance and collision analyses by commercial and military operators alike, aimed at long-term sustainability of the space environment. SSA's network of sensors has added a new leaf to the enhanced capability in surveillance with the launch of the first of the constellation of the US' Space-Based Space Surveillance (SBSS) satellites in September 2012. The US military, the lead agency in the world on SSA, has provided 'close approach' warnings averaging 150 a year to operators of over 1,000 operational satellites.[22] In the realm of military concerns, such a capability finds favour in 'keeping track' of the other active satellites too, satellites that could be employed in hostile counter-space options in times of need. The activities range from detecting a space launch, to monitoring what goes into orbit and cataloguing the activities of these objects for the duration of their operational lives. Such a capability would aim to analyse satellite or threat characteristics, capabilities, intent and current disposition, to eventually arrive at timely and actionable intelligence to negate the threat.

Ground Control Stations: Ground support infrastructure, so vital for the utility of the satellite network, represents the most vulnerable segment[23] of the space system because it could be subject to attack by a variety of means. The vulnerability increases when viewed in the context that even non-state terrorist organisations and traditional state entities that do not possess space-launch capability for ASATs, could find targeting ground stations a more viable and cost-effective option. Establishing these stations in remote locations with adequate dispersion and built-in redundancy does provide a certain degree of protection, which could further be increased by operationalising mobile control stations, especially for some pre-identified critical and sensitive satellite networks.

Autonomous Operations: In the eventuality of loss of a ground control station, satellites must retain the capability to perform certain autonomous operations like orbit and momentum control, enabling the satellite to function through the degraded control and telemetry communications links. Building such satellites with separate, independent-control systems would increase their ability to carry on their mission. Coupled with an onboard system for attack reporting, the satellite then could be more effective for system survivability.

System Hardening: Given the costs of launching satellites into orbit, contemporary satellite designs have focussed on squeezing the most utility out of each kilogram sent to space – the most 'bang for the buck'. Of late, there has been added interest shown in hardening of satellites' components as hardening of the space system's elements is the single most effective survivability measure.[24] Hardening of electronic components protects them from damage against the natural space environment as well as from the effects of nuclear weapons. Reflective surfaces and radar-absorbing materials reduce the 'visibility' of a satellite to sensors while appropriate geometry in the design of the spacecraft, aims to diminish exposed satellite surfaces and lessens the probability of penetration of counter-space tools.[25] Nano-materials are being explored for their possible use as protection against Direct Energy Weapons (DEWs), electromagnetic interference and radiation hazards. These protective features on offer by technologies of the future, of course, will come with the associated penalty of additional costs.

Proliferation of Satellites: This concept advocates placing multiple satellites in an orbit with overlapping coverages, with the premise that if one satellite is targeted, other satellites will still continue to execute the essential mission objectives. As a corollary, the concept would work equally well by the capability of launch-on-demand – of launching a replacement satellite on short notice. The adversary would be forced to attack multiple satellites to degrade the network, driving up the complexity and cost of such an operation. Technologies for further micro-miniaturisation of satellites and their components will enhance the efficacy and cost-effectiveness of such a proliferation strategy.[26]

Satellite Manoeuvre: An action to manoeuvre a satellite away from an inbound threat can only materialise on timely detection and assessment of the threat. Onboard thrusters are designed for station keeping and orbit control. Thrusters designed to take the satellite away from a collision course would need to carry additional propellant and be heavier, adding to weight penalties for the system. Such a manoeuvrability activity has been successfully undertaken several times in recent years for satellites as well for the ISS. The issue could also be addressed if technology can be honed to refuel a satellite in orbit, work on which is underway by the US Defence Advanced Research Projects Agency's (DARPA's) Orbital Expressway Programme.[27]

Anti-Counter-Space Operations: While the actions listed above are defensive and partly reactive in nature, there are options galore for active hard-kill operations that aim to destroy the adversary's counter-space weapons. Timely warning permitting, ground-based DEWs or high-power laser could intercept hostile direct-ascent, kinetic-energy missiles, destroying or rendering them ineffective before they reach their intended targets. Bodyguard satellites could be satellites co-orbiting with high-value satellites, which could position themselves between the satellite and the incoming threat when discerned. Space-based anti-ASAT platforms or kinetic-kill systems, more technologically feasible than surface-based directed-energy weapons, could be employed to intercept an attacking ASAT system and destroy it prior to its reaching the targeted satellite. Going a step further, a satellite itself could carry a small onboard self-defence missile system to increase its survivability.

Space would continue to remain a hostile environment as nature has designed it to be. Threats to spacecraft from space radiation and cosmic bodies would continue to challenge the scientific community in protecting their space-based assets. As more and more countries join the space bandwagon, in terms of both space launches and satellites in orbit, and more satellites are added to the fleet that is already in orbit by the current spacefaring nations, the proliferation of man-made objects in space will increase exponentially in the years to come. The risks and threats to satellites from these new additions might not yet be significant enough to cause alarm but what would definitely be of concern will be accidental collisions with space debris, an event that would unleash long-term cascading effects.

Three countries have already tested ASATs in space. As more countries hone the technological expertise in space launchers, spinoffs in the realm of ASAT capabilities are bound to be a natural fallout. The threat of ASATs looms not just from countries that possess space launch capability but equally so from countries armed with ballistic missiles. Over 40 countries possess the capability of operating micro-satellites, satellites which have the inherent potential of being used as anti-satellite weapons. Technologies that are emerging today in the field of lasers and beam weapons will only serve to add to the challenging complexity of safeguarding space assets.

Survivability of space assets, thus, would need to be addressed right from the satellites' design stage. Technologically, added impetus to research and development efforts would be a key ingredient of anti-counter-space operations that aim to reduce the vulnerability, and increase the sustainability, of spacecraft in any future conflict. At the global level, it would serve all the stakeholders well, regardless of nationality, to collaborate in the field of SSA and merge into a global SSA network that contributes jointly in making space a safer haven for the benefit of humanity.

Notes

1. "Russian and US Satellites Collide," BBC News, February 12, 2009, accessed at http://news.bbc.co.uk/2/hi/ 7885051.stm, on December 4, 2012.
2. Carl Hoffman, "China's Space Threat: How Missiles Could Target US Satellites," December 18, 2009, accessed at http://www.popularmechanics.com/technology/military/satellites/4218443?click=main_sr on December 7, 2012.
3. NASA USA, "WISE Revises Numbers of Asteroids Near Earth," accessed at http://www.nasa.gov/mission_pages /WISE/multimedia/gallery/neowise/pia14734.html, on December 17, 2012.
4. Joseph A. Angelo, Jr, *Encyclopedia of Space and Astronomy* (New Delhi: Viva Books Private Limited, 2010), p. 58.
5. Ibid., pp. 395-396.
6. Carolyn Fry, "CO2 Prolongs Life of Space Junk," BBC News, May 5, 2005, accessed at http://news.bbc.co.uk/2/hi/ science/nature/4486049.stm, on December 17, 20012.
7. Paul Marks, "Satellite Collision 'More Powerful than China's ASAT Test," February 13, 2009, accessed at http://www.newscientist.com/ article/dn16604-satellite-collision-more-powerful-than-chinas-asat-test.html, on December 18, 2012.
8. Stuart Clarke, "How Dangerous is Space Debris?," *The Guardian*, October 4, 2012, accessed at http://www. guardian.co.uk/science/across-the-universe/2012/oct/04/astronomy-space, on December 18, 2012.
9. "Chinese ASAT Test," June 22, 2012, accessed at http://www.celestrak.com/events/asat.asp on December 18, 2012.
10. "Recent US and Chinese Antisatellite Activities," (USA) *Air & Space Power Journal*, Fall 2009, accessed at http:// www.airpower.au.af.mil/airchronicles/apj/apj09/fal09/mackey.html on December 20, 2012.
11. NASA, "The Threat of Orbital Debris," accessed at http://images.spaceref.com/news/2009/ODMediaBriefing 28Apr09-1.pdf on December 18, 2012.
12. Clarke, n.8.
13. Angelo, n. 4, p. 558.
14. Leonard David, "Russian Satellite Hit by Debris from Chinese Anti-Satellite Test," March 8, 2013, accessed at http://www.space.com/20138-russian-satellite-chinese-space-junk.html on March 11, 2013.

15. Matt Ford, "Orbiting Space Junk Heightens Risk of Satellite Catastrophes," February 27, 2009, accessed at http://arstechnica.com/science/2009/02/orbiting-space-junk-heightens-risk-of-satellite-catastrophes/ on December 21, 2012.
16. Micah Zenko, "A Code of Conduct for Outer Space," accessed at http://www.cfr.org/space/code-conduct-outer-space/p26556 on March 18, 2013.
17. UN OOSA, "Long-Term Sustainability of Outer Space Activities," February 8, 2010, at OOSA's 47th session at Vienna, February 8-19, 2010, accessed at http://www.unoosa.org/pdf/limited/c1/AC105_C1_2010_CRP03E.pdf on Mar 18, 2013.
18. Francis Harris, "Beijing Secretly Fires Lasers to Disable US Satellites," September 26, 2006, *The Telegraph,* June 27, 2013, accessed at http://www.telegraph.co.uk/news/worldnews/1529864/Beijing-secretly-fires-lasers-to-disable-US-satellites.html on January 24, 2013.
19. USA's Defence Science Board, "Directed Energy Weapons," December 2007, accessed at http://www. acq.osd.mil/dsb/reports/ADA476320.pdf, on January 24, 2013.
20. Col. James G. Lee, "Counterspace Operations for Information Dominance," in Col. Bruce M. DeBlois, ed., *Beyond the Paths of Heaven: The Emergence of Space Power Thought* (Maxwell AFB, AL: Air University Press, 1999), p. 284.
21. Rahul Prakash, "China and Cyberspace," *ORF Issue Brief*, no. 52, March 2013.
22. Brian Weeden, "Space Situational Awareness Bigger Than U.S. Military," September 30, 2012, *Defense News*, accessed at http://www.defensenews.com/article/20120930/DEFFEAT05/309300008/Space-Situational-Awareness-Bigger-Than-U-S-Military, on January 29, 2013.
23. Majors Brian Garino and Jane Gibson, USAF, "Space System Threats," accessed at http://space.au.af.mil/au-18-2009/au-18_chap21.pdf on January 14, 2013.
24. Tom Wilson, "Threats to United States Space Capabilities," accessed at http://www.fas.org/spp/eprint/article05. html#17 on January 31, 2013.
25. Lt. Col. James Mackey, USAF, "Recent US and Chinese Antisatellite Activities," (US) *Air & Space Power Journal*, Fall 2009, accessed at http://www.airpower.au.af.mil/airchronicles/apj/apj09/fal09/mackey.html ojn January 31, 2013.
26. Wilson, n. 24.
27. Ibid.

6

The Laws of Space

Laws are spider webs through which the big flies pass and the little ones get caught.

— Honore de Balzac
French novelist, 1799-1850

January 11, 2007 witnessed the Chinese conduct an ASAT test which involved destruction of a defunct Chinese weather satellite at an altitude of 865 km by a ground-launched missile interceptor. The event prompted a global outcry as the world reacted with alarm and dismay, partly because China remained silent on this event till almost two weeks after the incident and partly due to concerns that the test would herald a space race. More condemnable was the fact that the destruction of the 750 kg satellite created thousands of pieces of debris of sizes that are even now potentially a threat to other orbiting space assets – debris that will continue to reside in the LEO for decades to come. The debris-creation, at worst, may be regarded as an act of irresponsibility by a nation-state but the fact is that this singular space activity was not in violation of any existing space law or legislation, except for a feeble concern for maintenance of the space environment enshrined in the Outer Space Treaty, 1967. The US knew in advance of this event but chose to say nothing till after the test.[1] Could the reason be that there are no laws that mandate prevention of such an occurrence?

China had been in the know of global concerns about debris, as its 'White Paper on Space, 2006' chronicles China's "active participation" in the Inter-Agency Space Debris Coordination Committee and documents initiation of a Space Debris Plan.[2] Being aware of the hazards that such debris could pose, including to space assets of its own, were the Chinese motives to go ahead with the test beyond just heralding the arrival of their ASAT capability? Or was it a predictable and unfortunate response to the US' space policies[3] – policies of declining to negotiate an agreement governing the 'peaceful' uses of space proposed by Russia and China in 2002 and again in 2008, while going ahead with domestic advocacies of space "dominance" and "control" in its national space policies?

The two examples quoted above arising out of a single space activity bring to the fore that international agreements and treaties concerning the use of space are obviously not adequate enough to provide safety and security to space assets. Almost 60 countries have assets in space that contribute to the larger economic and social good of the countries concerned. Some of these countries have stakes that are in the realm of the military, including assets that are of dual-use. As more countries join the space bandwagon there is every likelihood of rivalries and clash of interests between nations as they compete for a slice of the space pie. It would be stating the obvious that it is only a coordinating body at the global level that can have a mechanism in place that oversees and resolves such conflicts that may arise between conflicting member states. For this body to continue to be of relevance, its enforcing mechanisms need to be binding on the said states. This chapter throws light on some of the important international agreements, including key areas where they are found wanting and suggests some 'rules of the road' that could be addressed to plug the gaps in the existing framework.

The Existing Legal Framework

International Treaties and Agreements

The space age commenced with the launch of the satellite Sputnik by the Soviet Union in October 1957. This was the era of the Cold War between the two superpowers, each competing with the other to achieve 'firsts' in space, the fourth dimension. Where went the explorer, there followed the lawyer. Born out of anxiety about the Cold War and viewing the potential of space as a common heritage of mankind, a debate began in the UN on mechanisms to ensure the peaceful use of space that led to creation of the UN Committee on the Peaceful Uses of Outer Space (UNCOPUOS) in 1959. By 1979, five international treaties and agreements, as part of Space Law – the body of law applicable to, and governing, space related activities – were formulated and in force. UNCOPUOS has 74 member-states as of January 2013.4

Outer Space Treaty, 1967: The *Treaty on Principles Governing the Activities of States in the Exploration and Use of Outer Space, including the Moon and Other Celestial Bodies,* commonly referred to as the Outer Space Treaty (OST), forms the basic framework of international space law and is the most widely-adopted treaty, with 102 member states as of January 2013.[5] The governing principles of the treaty lay down for its member states the norms of use of outer space – that its exploration and use is free for all states and shall be carried out for the benefit of humanity and in the interests of all countries; that outer space is not subject to national appropriation by claim of sovereignty, by means of use or occupation; states shall not place nuclear weapons or other weapons of mass destruction in orbit or on celestial bodies; the Moon and other celestial bodies shall be used exclusively for peaceful purposes; astronauts shall be regarded as the envoys of mankind; responsibility of national space activities whether carried out by governmental or non-governmental entities rests with the state which will also be liable for damage caused by their spacecraft and states shall avoid harmful contamination of space and celestial bodies. The Partial Test Ban Treaty of 1963, though not a space law, as a forethought had already banned testing of nuclear weapons in space.

The Rescue Agreement, 1968: The *Agreement on the Rescue of Astronauts, the Return of Astronauts and the Return of Objects Launched into Outer Space*, or the Rescue Agreement, elaborating and giving further concrete expression to the elements of the Outer Space Treaty, provides that states shall take all possible steps to rescue and assist astronauts in distress and promptly return them to the launching state, and that states shall, upon request, provide assistance to launching states in recovering space objects that return to the Earth outside the territory of the launching state. It also requires that any state on being aware of any personnel of a spacecraft in distress, must notify the launching state and the UN. If the distress occurs in an area outside the territory of the launching state, then any other state that is in a position to, shall extend assistance in the rescue. Any state that has any space object landing on its territory, at the request of the launching state, shall recover and return the object and be compensated the costs incurred thereupon.

The Liability Convention, 1972: The *Convention on International Liability for Damage Caused by Space Objects,* or the Liability Convention, brings to bear responsibility on a state for all objects launched from its territory into space and is fully liable for damages that may arise from that object. Two states launching an object jointly are both liable for the damage. This convention applies at the level of states only, that is, if an individual suffers injury or damage, then the individual is to arrange through his country the claim for compensation from the launching state. The convention also provides for procedures for the settlement of claims for damages.

The Registration Convention, 1975: *Convention on Registration of Objects Launched into Outer Space,* commonly called the Registration Convention, requires states to furnish to the UN data about each space object to include the name of the launching country, designation/registration number, date and location of launch, basic orbital parameters and the general purpose of the space object. The register is kept at the UN Office for Outer Space Affairs (UN OOSA) and as of January 2011, has details of 93.5 percent of all functional space objects.[6] A resolution in 2007 recommended that the details of space objects should also include the Coordinated Universal Time as the time reference for launch, GEO location where applicable, change in the status in operation of spacecraft, change in ownership, approximate date of decay or reentry, date and physical conditions for moving the object to a disposal orbit and web links to official information on the space object.

The Moon Agreement, 1979: *Agreement Governing the Activities of States on the Moon and Other Celestial Bodies,* or the Moon Agreement, provides that the Moon and other celestial bodies are the common heritage of mankind and activities on such bodies are to be governed by an international regime. The agreement duly recognising the potential that the Moon offers in terms of exploitation of natural resources and for furthering exploration of outer space, desires to prevent the Moon from becoming a source of international conflict. It provides for activities on the Moon purely for peaceful purposes, banning military use, including weapon testing or as a military base; allows for exploration and other celestial activities with the approval of the UN and in so doing, declares the desirability of making samples available to all countries/scientific communities for research; bans altering the environment of celestial bodies and requires that states undertake measures to prevent accidental contamination; bans any state claiming sovereignty over any territory and allows freedom to all to undertake scientific investigation on the basis of equality, including setting up manned or unmanned stations.

Legal Principles

Over and above the treaties and agreements listed above, space law also includes international principles, conventions, rules and regulations of international organisations (e.g. the International Telecommunications Union), national laws, rules and regulations, executive and administrative orders, and judicial decisions.[7] Amongst them there are five sets of principles which have the legal status of General Assembly Resolutions that provide generally accepted principles, rules, responsibilities and standards by which states may, and very often do, govern their space related activities.

The Declaration of Legal Principles Governing the Activities of States in the Exploration and Uses of Outer Space, 1963: This declaration offers guiding principles in the exploration and use of outer space, in that, it is to be carried out for the benefit, and in the interests, of all mankind; no nation may claim sovereignty; activities of states are to be carried out in accordance with international law; nations undertaking such activities must accept responsibility for the governmental or non-governmental agency involved and are liable for any damage caused, and states shall be guided by the principle of cooperation and mutual assistance and shall conduct all their activities in outer space with due regard for the corresponding interests of other states.

The Principles Governing the Use by States of Artificial Earth Satellites for International Direct Television Broadcasting, 1982: Every state has an equal right to conduct activities in the field of international direct television broadcasting by satellites and to authorise such activities by persons and entities under its jurisdiction and such activities shall be carried out in a manner compatible with the sovereign rights of states, including the principle of non-intervention. Such activities are to be based on, and encourage, international cooperation.

The Principles Relating to Remote Sensing of the Earth from Outer Space, 1986: Fifteen principles are listed under this category. Remote sensing activities shall be carried out for the benefit, and in the interest, of all countries and conducted in accordance with international law including the relevant instruments of the International Telecommunication Union. Such activities shall not be conducted in a manner detrimental to the legitimate rights and interests of the sensed state but shall promote international cooperation by making available technical assistance to other interested states on mutually agreed terms. States conducting remote sensing activities that have identified information in their possession that is capable of averting a natural disaster or any phenomenon harmful to the Earth's natural environment shall disclose such information to the states concerned.

The Principles Relevant to the Use of Nuclear Power Sources in Outer Space, 1992: Activities involving the use of nuclear power sources in outer space shall be carried out in accordance with international law. The principle lists guidelines and criteria for safe use of nuclear energy by mandating its use only on those missions which cannot be operated by non-nuclear energy in a reasonable way and that too after conducting a thorough and comprehensive safety assessment prior to launch. Such states shall endeavour to protect individuals, populations and the biosphere against radiological hazards. Nuclear reactors may be operated on interplanetary missions, in sufficiently high orbits or in LEO, provided they are parked in sufficiently high orbits after the operational life of their mission. The launching state shall inform the UN and the states concerned in the event the space object malfunctions, with a risk of reentry of radioactive materials to the Earth.

The Declaration on International Cooperation in the Exploration and Use of Outer Space for the Benefit and in the Interest of All States, Taking into Particular Account the Needs of Developing Countries, 1996: States are free to determine all aspects of their participation in international cooperation in the exploration and use of outer space on an equitable and mutually acceptable basis and in so doing, particular attention should be given to the benefit and the interests of developing countries and countries with incipient space programmes stemming from such international cooperation.

Other Significant Agreements

Besides the ten treaties and principles covered above, the UN General Assembly has passed 104 other resolutions relating to outer space, as of January 2012. All these resolutions have aimed at long-term sustainability of space – to make outer space a safer and weapons-free environment – in the backdrop of international cooperation. A large number of these resolutions has been adopted without a vote. Of the balance that were adopted by a vote, interestingly, the USA has voted against, or abstained from voting in, almost all of them, with Israel a close runner-up. Amongst the other prominent spacefaring nations – those with independent launch capability, thus, giving them the status of "launching states" as enshrined in the UN space legislation – the Russian Federation, China, France, Japan and India have largely voted for the resolutions.[8]

International Telecommunication Union (ITU)

The ITU is the leading UN agency for information and communication technologies. As the global focal point for governments and the private sector, the ITU allocates global radio spectrum and geostationary satellite orbits, and develops the technical standards for seamlessly interconnecting networks and technologies. Based in Geneva, Switzerland, the ITU has a membership of 193 countries and over 700 private sector entities, including academic.[9] Ratification of ITU Conventions implies acceptance by member entities, regulations of which are legally binding on the signatories.

The ITU was founded initially in Paris in 1865 as the International Telegraph Union. It took its present name in 1934 and in 1947, became a specialised agency of the United Nations. Although its first area of expertise was the telegraph, the work of the ITU now globally covers the whole Information and Communication Technology (ICT) sector, from digital broadcasting to the Internet, from mobile technologies to 3D TV, cyber security and monitoring of climate change. The ITU operates as a coordinating agency but is vested with no enforcing powers or mechanisms. In the realm of orbiting platforms, the work holds equally true in allocation and coordination of frequencies for communications, Earth observation and science spacecraft, and in any other spacefaring craft that communicates using radio frequencies. Under the auspices of the UN, the ITU administers the radio frequency spectrum worldwide, in a manner that the needs of all member nations and agencies are met, in terms of avoidance of interference, and resolution of potential disputes and conflicts. Towards this end, the ITU has divided the world into three ITU regions for the purposes of managing the global frequency spectrum. Each region has its own set of frequency allocations. The aim is not only to view the frequency spectrum as a limited natural resource that must be used rationally, efficiently and economically, but also that countries have equitable access to it.

An equally important role carried out by the ITU in matters of space is of allotting GEO orbital slots. All member satellite operators are mandated to register with the ITU for both orbital position and frequency before launching GEO satellites. It is also mandated that commercial satellite operations are duly licensed by a national administration prior to operation. However, military frequencies are exempt from this regulation other than the requirement to ensure that the military frequencies in use do not cause interference to other users.[10] The allocation is made by the ITU on a "first come, first served" basis. Once made, the member state has a fixed time to occupy the allocated orbital position, as failure to do so would invite re-allocation to another bidder.[11] Once registered, the approved slot and the frequency are deemed to be in effect until the operator-defined system's life expectancy has expired, or no longer in use under due notification to the ITU. The ITU has, by design, no enforcement powers or authority. The member states choose voluntarily to abide by the rules and regulations enunciated by the ITU and work to resolve any conflict about spectrum usage in good faith and considerations of mutual benefit.[12] The ITU remains a coordinating agency for the international use of the frequency spectrum, while the authority to launch and place a satellite in orbit and operate on the given frequencies, rests with the concerned sovereign state. At the sovereign or member state level, the state's administration is the link between the ITU and satellite operator. The state administration, as the domestic regulatory body, participates in the ITU activities and oversees the assignment and use of frequencies within the state's domestic borders. At the planned user end, the user proposing to launch a GEO satellite, applies for the orbit and frequency to the domestic regulatory body, which, in turn, conducts an internal coordination and applies to the ITU for registration five to two years prior to launch of operation, specifying details of channel frequencies, orbital position, geographical location of the intended Earth station, transmitter power and data format.

The ITU has also addressed the needs of countries which to date do not possess the capability of placing satellites in GEO but may attain the same in the years to come. Under the principle of right to equitable access to resources, the ITU established *a-priori* plans in which a certain amount of frequency spectrum and GEO orbit slots are set aside for future use by such countries.

Inter-Agency Space Debris Coordination Committee (IADC)

The IADC is a forum of international and national space agencies for the worldwide coordination of activities concerning man-made and natural debris in space. The member states include almost all the important launch-capable countries and the ESA. The primary purpose is to exchange information, facilitate cooperation and explore options for debris mitigation. The IADC formulated Space Debris Mitigation Guidelines that amongst others, listed specific guidelines to states to move end-of-mission GEO satellites to graveyard orbits 235 km above GEO and remove satellites from LEO within 25 years. UNCOPUOS in its last session on IADC in February 2013 tabled a study report of the IADC that addressed the LEO band and acknowledged the altitude band between 800 and 1,000 km as the most dense and dangerous,[13] a situation that has reached a 'tipping point'.

India is a member of UNCOPUOS while its space organisation, ISRO, is a member of IDAC. ISRO addresses debris mitigation issues in its launchers by way of designing them in a manner that avoids their breaking up at the end of its mission. In addition, the last stage being a liquid-propelled one, the ejecta does not contain any solid matter that could add to space debris. Indian geostationary satellites cater for additional propellant to reorbit them into a higher 'graveyard' orbit at the end of their operational life. India has been found to be contributing significantly to the international activities in the field of space debris.[14]

National Laws

The international space law that defines space as a 'common heritage of mankind', mandates that countries partaking in national activities in space bear international responsibility. To ensure that such activities are in conformity with the international legal regime, some countries have enacted their own domestic space law or legislation. National space legislation that has been formulated and brought in vogue by independent countries, thus, aims to govern and guide the country's national space activities, including the activities of non-governmental entities such as commercial and non-profit organisations.

Amongst the many reasons that countries adopt national law, two stand out significantly. The first and foremost is the country's concern for its national security. With more and more assets of military value being placed in orbit, including satellites of dual-use, the nation's security concerns play a predominant role as against satellites purely for peaceful purposes. The need, thus, is of domestic legislation that the government enacts to prioritise furtherance of its space goals. The second factor that encourages advocacy of such a law is the increasing role that the private and commercial sector has brought into play in pursuance of its commercial interests. International law requires the "launching state" to bear responsibility for its ventures even when private players are involved. It is then with reason that the USA, the Russian Federation, China, the UK, France, Japan and South Africa are some of the countries that have a national space policy or law in place.[15]

National Law in the Indian Context

India's space odyssey spans five decades from the time it launched its early rockets in the 1960s. Since then, India has been an active participant in almost all fora of international space legislation. It is a signatory to the Moon Agreement and has ratified all the other four UN treaties and agreements. Two other significant treaties ratified by India are ones that ban nuclear weapons tests and the use of the military in outer space. Not only are there scores of Indian satellites in LEO and GEO, but India is also one amongst only eleven countries in the world to possess an indigenous space-launch capability, a capability that has opened global opportunities enabling the country to venture into the commercial launch sector. It has been the fourth country to land a spacecraft on the Moon. The government led space programme has matured over the years inviting public-private partnerships which began to play an increasing role in the space programme.

With these wide ranging activities encompassed in the Indian space programme and the global acknowledgement of India being a mature and responsible space power, it is indeed rather surprising that the nation has no national space policy or law. The Indian space programme is guided by the Allocation of Business Rules for the Department of Space (DoS), along with related legislations and regulations of the government and individual policies on remote-sensing data, satellite communications, mapping, etc.[16] What remains lacking is a comprehensive national space legislation or a specific law that deals with space activities undertaken by the country in the contemporary context, which could also factor in activities in the years to come. The reason for not legislating such a domestic law could well be the fact that the Constitution was adopted in 1950, well before forays into space that commenced in the 1960s. The country is considered to be slow in ratifying or acceding to treaties and takes its own time to incorporate its convention obligations into domestic law.[17] Some are also of the view that the space activities till 2000 were purely in the government's domain with no private partnership and the same did not necessitate articulation of a domestic space law.[18] Is it time now to relook at the question: does the nation need to formulate and adopt a national space policy, from which then should flow the national space laws? Some factors that offer significant insight into the urgent need of such legislation are briefly discussed in the following paragraphs.

Ever-increasing Participation of the Private Sector: Commencing in the early 2000s, the domestic space sector has been opened up for participation by commercial private enterprises. Their role, as guided by Antrix, the commercial arm of ISRO, grew significantly with over 500 companies participating in the space effort,[19] with almost 60 per cent of launch vehicle cost flowing back to the industry.[20] Launch services apart, proliferation of civilian commercial applications in the space sector necessitates a larger involvement of the private players. With the entry of private entities into satellite systems, including licensing, it is the Government of India that bears international responsibility and liability. This, in turn, dictates the need to adopt at the national level, an unambiguous, transparent and user-friendly legal regime supported by readily available information for the private enterprise – a legislation that not only regulates and governs but also has the force of law over all the stakeholders involved in the national space sector.

Operation of Multiple Agencies: Information on the space programme, its aims and objectives, and administrative guidelines is available largely from the websites of ISRO and Antrix.[21] The website of the Department of Space[22] lists individual policies on satellite communications and remote sensing data. At the apex level, the inter-ministerial Space Commission coordinates space-related issues for policy-making and implementation through the Department of Space and ISRO. Applications that are served by space-based platforms have inter-linkages with governmental agencies like telecommunications, meteorology, education, home affairs, natural resources, etc. and, over the last decade, have included private enterprises in a big way. It is the Space Commission that needs to regulate the interdependence of each of the stakeholders under a common and transparent legal framework which then could be binding on all concerned through well conceived legislation at the national level.

Global Launch Services: ISRO ventured into the field of providing commercial launch services to its foreign partners in 1999. Due to its relative low cost and robust reliability, the PSLV remains a very viable option for launching satellites into LEO, having orbited 35 foreign satellites so far.[23] For all such customers worldwide, India bears responsibility and liability as the "launching state" as enshrined in UNCOPUOS' Rescue Agreement, Liability Convention, and Registration Convention. While India, having ratified all the three treaties, bears responsibility at the UN level, it would serve the country well to include in its national space law certain standardised operating protocols that identify responsibility that ought to be vested with the foreign partner, specially with respect to dispute resolution, debris mitigation, insurance and financial liability.

Safeguarding Intellectual Property (IP) Rights: The increasing role that the industry has come to symbolise in the domestic space activities under the aegis of Antrix, has led to the transfer of some space technologies, including spinoffs, to private entities in the fields of communications, computers, electronics, optical and remote sensing data utilisation equipment. The protection of Intellectual Property (IP) rights is, thus, a natural corollary to safeguard the interests of development agencies and innovators, which are carried out by ISRO by way of patents, copyrights, trademarks and data rights. Equally true in the field of remote sensing data that is marketed, both nationally and internationally, no information is known regarding the terms and conditions of such marketing contracts.[24] It would be a corollary that such safeguards on IP rights, which otherwise are covered under national insurance regulations, need to be looked into separately within the ambit of the space policy and be an essential constituent of the Indian space law.

Activities in Support of the Military: The genesis of the Indian space programme has civil-oriented roots and the focus has remained so for the past five decades. The three Services of the military have been relying on some dual-use satellites for imaging and communication purposes, albeit to a limited extent. The talk of a Space Command for the three wings of the military has been doing the rounds for the last decade but the roots of this organisation are yet to take shape. Given China's lead in activities in space, especially in the realm of military capabilities and the geostrategic environment that India finds itself in, it would serve India well to review its security paradigm and look at space as yet another medium that can contribute to the country's national security. While the issue of space-based assets contributing to national security *per se* lies beyond the purview of national space law, a law in place, would facilitate dovetailing military-space requirements into the national space effort. The national policy should also address the issue of the export control regime to prevent proliferation of dual-use and launch technologies that could power ballistic missiles, while offering launch services commercially.[25] Of special significance should be a law that could address the issue of need-based prioritisation of launch and satellite construction in support of the military.

International Space Law and its Flaws

Adopted over four decades ago at the peak of the Cold War, the international space law has been based on the centrality of the *res communis* principle, i.e.

space is viewed as a global common and belongs to mankind and not to an individual country or entity. Over time, the emerging space technologies and capabilities, and the contemporary socio-political and economic world order bring to the fore the fact that space law today lacks adequacy and adequate mechanisms for conflict resolution and no longer has the far-reaching relevance that is warranted by the growing number of participants, including commercial entities.

Treaties and agreements applicable for member states are adopted through resolutions of the UN's General Assembly (GA). These resolutions do not lay down rules but state the position of the legislature and generally being procedural resolutions, are non-binding on the states (except on matters that pertain to payments due to the UN). The UN remains a policy-charting organisation devoid of mechanisms to enforce those policies. The implementation of policy recommendations flowing out of these resolutions and decisions thereafter rests with the individual member states. Responsibility by a member state towards discharge of such resolutions, thus, "is assumed", with no mechanisms of compliance or enforcement vested with the UN. The decision-making process within COPUOS is cumbersome and slow, further compounded by the fact that a small minority of states seems to exercise a kind of veto power as they have taken over the decision-making process, using the requirement of consensus.[26]

The legalese of the agreements is such that it is open to broad interpretation, allowing signatories to view them from an angle that suits their perspective. The language of the resolutions has been found to be inclined towards the ambiguous. There exist loopholes and a general lack of explanatory notes. The space treaties in particular contain terms that lack definition and appear nuanced and ambivalent. While laying down exhaustive space treaties and agreements, space *per se*, is not even defined. Delineation of the boundary between air space, where national sovereignty is a given, and space, the global common that is free for exploration by all, has never been attempted. The term 'peaceful uses of space' is subject to varied interpretations considering that a large majority of space assets is of dual-use nature. Given the special status of astronauts, would a commercial space traveller qualify for a similar definition? If yes, would the status of 'envoys of mankind' be warranted, nay justified, for a billionaire who pays for his space-ticket? The Rescue Agreement, similarly, does not specifically cover space tourists. The Liability Convention provides for fault-based liability and responsibility for damage arising out of debris collision. How is fault to be determined if a large number of debris pieces remain untracked and their source of origin unknown?

Could the reason for couching such terms in diplomatic nuances be that it is easier and non-contentious to achieve consensus and majority voting if the language and terms are vague and not bound to legally-sustainable definitions? Lack of such clear-cut definitions prompted the introduction of the term "space flight participants" with respect to crews to the International Space Station (ISS), in the Intergovernmental Agreement and MoU between participating nations.

The space law came about due to circumstantial necessity dictated by the Cold War. Other member states that joined the treaties did so when they had yet to achieve even a fraction of space capability enjoyed by the superpowers and, thus, remained largely bereft of the negotiation process, leading to adoption of treaties that addressed immediate concerns of the two superpowers with no futuristic perspective. Under the provisions of the Liability Convention, the "launching state" bears responsibility and liability of anything launched from its territorial sovereignty. A growing number of commercial entities today undertake launch operations, with a situation that gets compounded when a satellite operator is registered in one state, carries out operations for another and launches through a commercial entity from the soil of some other "launching state".

The Moon Agreement for one was designed to extend the reach of *res communis* so that individual states or private entities could not take advantage of the Outer Space Treaty's shortcomings and start staking claims to extra-terrestrial real estate. This treaty is considered to be a failed treaty as only 15 countries have ratified it as of January 2013. None of the countries that has sent or plans to send missions to the Moon are amongst these fifteen. India is one that has not ratified but only signed the said treaty. One predominant factor emerges to explain the relative lack of popularity of this treaty. The "common heritage of mankind" clause permits resource exploitation from the Moon but places a caveat that not only would such exploitation be monitored by an international regime but the resources extracted and the means of acquiring the same are to be shared on a equitable basis with those that do not possess this capability. Most argue that such a mandate inhibits states from venturing into such a domain – of funding a mission at the national expense and then being made to share the 'spoils of war' with the have-nots. During the time of adopting this agreement, the US and the Soviet Union were the only two nations that had the possibility of acquiring the capability of extra-terrestrial exploitation in the foreseeable future. The law of equitable sharing would remain impotent as long as the means to carry out such exploitation were in the hands of the powers that did not ratify the agreement. The US rejected it on the ground that unless the state had a right to benefit from exploitation, it would have no incentive to pursue such a venture.[27] The consequences turned out even worse for the flourishing commercial sector worldwide which has estimated annual revenues in excess of $200 billion in the space domain as of 2012.[28] The private sector would obviously be loath to enter such profit- or benefit-sharing ventures that inhibit its commercial pursuits. In any case, a meaningful and practical resource exploration of the Moon is at least two decades away as contemporary space technology needs to evolve drastically to make such a venture economically viable. It would well involve the efforts of multi-states and private conglomerates to fructify such ambitions. It would, thus, be in the global interests that Andrew G. Haley's words, "Law must precede man into space" be put into practice at least in matters of exploration of the Moon and other celestial bodies, by reshaping and honing the Moon Treaty in such a manner that it retains its relevance comprehensively in the years to come. The words would be equally prophetic in the field of space tourism – a reality on the horizon – an issue though not in the context of the Moon Agreement under discussion, but one which also remains unaddressed by the contemporary international law, necessitating an all-inclusive legal framework that addresses the role of state actors, commercial entities and private citizens, amongst others.

Another arena where the international space law regime is found lacking in clarity is that of the GEO. Lying 36,000 km above the Equator, the GEO has a finite number of 'parking slots' for satellites and, thus, is viewed as a finite resource. With 435 operational satellites already in position in GEO as of June 1, 2013,[29] the balance slots are fast filling up due to the versatility that such an orbit offers. Satellites being booked (with the ITU) and placed in orbit over the Equator belong to a few states. The 'common heritage of mankind' principle precludes equatorial states claiming sovereignty over the orbit, though the UN COPUOS recognises that by courtesy of their geographical location, such states be considered as having special rights to segments of GEO above their territory, to meet the ends of justice for those developing states that too aspire to gainfully exploit such orbits. A simple solution would be to earmark slots for such states and reserve them for their use[30] when technologically and economically viable for them in the future. ITU thereafter committed to earmark two GEO slots and associated frequencies to each of these states, regardless of other considerations.[31]

The functioning of the ITU took another hit as countries scrambled for the finite GEO slots. States, to further their national interests for the future, apply for registration of 'paper satellites', or non-existent satellites – an issue that had reached a point when out of 1,300 applications filed for registration, 1,200 were for paper satellites.[32] The ITU's mandate to operationalise a satellite within seven years of filing or lose the slot for failing to do so, has worked only marginally to reduce the coordination effort involved at the ITU's end. A similar predicament is faced on the aspect of frequency allocation for satellites. As of 1979, 90 per cent of the frequency spectrum was controlled by countries with only 10 per cent of the world's population. Bands below 30 GHz are particularly popular due to their inherent propagation characteristic of following the Earth's curvature that reduces the requirement of a number of relay stations. These relatively lower frequency bands comprise a scarce and contested resource as they accommodate a small number of users to avoid mutual interference.[33] Frequencies for military satellites lie outside the ambit of regulation by the ITU so long as the state caters for non-interference aspects. This too does not seem to bode well as the interference manifesting in satellite communications by the growing use of military frequency bands is on the rise.[34] The ITU, thus, lacks power and mechanisms towards enforcement and acknowledges that not all launching states report the correct or updated number of launches.

Management of debris is broadly covered by the OST and the guidelines enunciated by IADC. Some spacefaring nations have incorporated measures towards the same, including moving derelict satellites into safer orbits. For most other states, the means of adopting the guidelines are limited by financial and technological considerations. For the worldwide good, if a technologically-capable state was to undertake retrieval of a derelict satellite of a non-capable state, ownership of the satellite remaining that of the launching state, the two states would need an instrument specifically to address the legal contours of such a licensing arrangement, including, and very importantly, the cost involved. IPR issues too would come into play as technical details regarding such satellites would be essential to be conveyed from the owner-state to the retrieving-state.[35] The international space law so far remains silent on such issues and that has prevented any state from even initiating a debate on such purposeful mechanisms.

Deployment of weapons of mass destruction in space is banned under the provisions of the UN treaties. The same is not valid for weapons other than those for mass destruction and conveys the inadequacy of the existing legal framework in preventing weaponisation of space, an issue which is being dealt with at greater length in the succeeding chapter.

It would, thus, be stating the obvious that the contemporary international space law regime has shortcomings, both in the ambiguity of the legalese and in the detailed "rules of the road" for governing mankind's activities in space. To make the space regime more relevant, the need of the hour would be to contextualise the legal framework to make it at par with activities that are currently being undertaken and with those that are on the verge of being fielded. Towards a more participative response from the space-players, the regime needs to be vested with powers to enforce such legislation coupled with adequate verification mechanisms.

The European Initiative

With the playing field of space getting more and more congested as satellites continue to be orbited, space security and sustainability assume greater importance. Coupled with the fact that the UN space legislature does not quite measure up to 'doable' and 'enforceable' mechanisms that can step in for conflict management, the EU introduced the International Code of Conduct (CoC) in 2008 and submitted the revised draft to the UN COPUOS in Vienna in June 2012. The CoC is a voluntary and non-binding international instrument that provides "rules of the road" to spacefaring nations. Within the legal framework of the OST and other relevant treaties and agreements, the CoC aims to provide a complementary instrument that is guided by the following principles:

- Freedom for all to use outer space for peaceful purposes.
- Preservation of the security and integrity of space objects in orbit.
- Due consideration for the legitimate security and defence needs of states.

Arguments in Favour of the CoC

The International CoC is viewed largely not as an absolute framework that could address the shortcomings of the current legislation but one that would be a step in that direction. Proponents argue that the code would be a start to the process that brings states together to identify common concerns and to suggest possible remedies. It would also be a shared platform for exchanging information on SSA that would lead to added safety in space operations. Another point in favour would be in the field of preventing harmful activities in space. In that, in support of the provisions of the OST, the CoC would foster consultations and discussions amongst countries prior to the conduct of any harmful activities.[36] What the CoC aims to do is to set out voluntary best practices for nations to adopt towards breaking space security dilemmas. Facilitation of dialogue between countries would enable a better understanding of not just *what* countries are doing in space but the *how* and *why* of decisions that lead to such activities. The CoC, thus, could provide an opportunity for all countries to show their good intentions in their actions in space.[37]

The Arguments Against

The naysayers counter-argue that the effectiveness of the CoC, which is a soft law and not even a treaty, will remain a concern as its centrality lies on its non-binding and voluntary acceptance by countries. In fact, the CoC has no legally binding status even within the EU.[38] Lack of enforcement mechanisms will inhibit compliance by member states. Given its voluntary nature, the code will perforce rely on political and moral suasion to encourage safer space behaviour. While this will not prohibit harmful activities, reliance will need to be placed on imposing sanctions and costs on errant countries.[39] The general nature of the stated objectives in the code leaves considerable room for differing and even conflicting interpretations.[40] A shortcoming found in the present avatar of the CoC is also its inadequacy in addressing issues of the future. No cognisance seems to have been taken of the needs in the times to come. Towards global acceptance of the code, Li Jiqian argues that for the CoC to be effective, it has to get a favourable consensus from the leading spacefaring countries as the efficiency of a legal framework is a measure of acceptance by the key actors.[41] It would then be in the fitness of things to see how some of the key spacefaring countries have reacted to the CoC.

Response of Key Space Players

The EU's initiative has been received with mixed responses worldwide, ranging from support, to conditional support, to outright scepticism. Besides countries of the EU as proponents of the code, major space-players that support the code are Japan, Australia and Canada. The US, on the other hand, while welcoming the code for the promise of avoiding and mitigating debris globally, has decided to join with the EU and other countries to develop the international CoC but has made it clear that it "will not enter any code of conduct that in any way constrains (its) national security-related activities in space or (its) ability to protect the US and (its) allies".[42] Amongst countries that are not favourable to the code are China, Russia, India, Brazil and South Africa. Russia's and China's unwillingness towards support, in part, lies in their political posturing in preventing an arms race in space, by way of proposing a treaty on Prevention of the Placement of Weapons (PPWT) in outer space in 2008. The CoC falls way short of the intentions proposed in the PPWT and the two stand to lose substantial soft-power influence if a successful code is implemented.[43] China, in addition, has taken the stance that space debris should not be included in the CoC and also has objections on the code's insistence that states share information on their domestic national space policies, including objectives for security and defence related activities.[44] India and China hold similar views on non-acceptance due to their not being included in the formative process of the code. India, while maintaining its reservations on the non-binding status of the code, has welcomed the initiative in principle and thereafter been a part of the deliberations and consultations.

An analysis of the impact that the wide ranging responses received to the CoC suggests that the draft code is likely to be a failure unless other significant players like China, Russia and India too endorse it. This is evident from the fact that in the current geopolitical climate of waning Western hegemony amidst rising centres of power and influence in the Asian region, going ahead with an international process without the support of such key emerging players would lack legitimacy and fail to produce any tangible results. More so, as the code lacks any formal enforcement mechanisms, wilful compliance by states is valuable only if the major countries support the process.[45] Michael Listner further suggests that lack of participation by China and Russia could encourage other states to follow suit.[46] While all is not lost in the hope of an international framework that the CoC advocates, it would be a natural corollary to assume that for the code to find global acceptance, the concerns of the current naysayers need to be addressed holistically after due deliberations and consultations.

The field of space law is still in its infancy while the arena of space activities itself is on a trajectory of tremendous growth and ever enlarging potential. The stakeholders are currently besieged with the problems of ever increasing space debris and weapons in space, while the future beckons the prospects of space tourism and asteroid mining. High costs and not-yet-developed technologies for such futuristic exploits are not the only reasons for falling short of achieving such lofty ambitions. Voids in the contemporary space law have played their part in holding back mankind's greater leap forward.

Activities in space, as advocated by the treaties and agreements of the UN COPUOS, and the international CoC proposed by the EU, are themed on voluntary and responsible behaviour by member states. While states do so voluntarily, not much exists in real terms of mechanisms that can monitor, verify or enforce the country's obligations towards space security and sustainability. The issue gets added complexity as the number of stakeholders steadily rises. Not only do countries come together in their endeavours in space, participation by commercial entities and conglomerates adds a new dimension that highlights the importance of 'rules of the road' for the fourth dimension. Such rules come pre-packaged with political and legal ramifications that necessitate an all encompassing, transparent and enforceable legal framework that should aim at enabling and ensuring that each participant conducts his activities in space in a responsible manner with the underlying centrality of viewing and treating space as a global common – a common heritage of man.

Notes

1. Michael R. Gordon and David S. Cloud, "US Knew of China's Missile Test, but Kept Silent," *The New York Times,* April 23, 2007.
2. Shirley Kan, "China's Anti-Satellite Weapon Test," *CRS Report for (US) Congress,* April 23, 2007, accessed at http:// www.fas.org/sgp/crs/row/RS22652.pdf on February 11, 2013.
3. Mark Kaufman and Dafna Linzer, "China Criticized for Anti-Satellite Missile Test," *Washington Post,* January 19, 2007.
4. UNCOPUS, accessed at http://www.unoosa.org/oosa/en/COPUOS/copuos.html on February 20, 2013.
5. UN Office for Outer Space Affairs, "Status of International Agreements relating to Activities in Outer Space," accessed at http://www.unoosa.org/oosa/en/SpaceLaw/treatystatus/index.html on August 14, 2013.
6. UN OOSA, "Registration of Objects Launched into Outer Space," accessed at http://www.unoosa.org/oosa/ SORegister/regist.html on February 15, 2013.
7. UN, "Space Law: Frequently Asked Questions," accessed at http://www.unoosa.org/oosa/en/FAQ/splawfaq.html on February 20, 2013.
8. United Nations, "Index of Online General Assembly Resolutions Relating to Outer Space," accessed at http://www.unoosa.org/oosa/en/SpaceLaw/gares/index.html on February 21, 2013.
9. ITU, "Overview," accessed at http://www.itu.int/en/about/Pages/default.aspx on February 21, 2013.
10. Wg Cdr Kaza Lalitendra, *Militarisation of Space* (New Delhi: KW Publishers Pvt Ltd, 2010), p. 44.

11. Development, Concepts and Doctrine Centre (DCDC), Ministry of Defence, UK, *The UK Military Space Primer* (Wiltshire: DCDC, 2010), Ch. 2, pp. 2-7.
12. Sqn Ldr K.K. Nair, *Space: The Frontiers of Modern Defence* (New Delhi: Knowledge World, 2006), p. 104
13. UNCOPUOS, Report of an IADC Study "Stability of the Future LEO Environment," February 11-22, 2013, accessed at http://www.oosa.unvienna.org/pdf/pres/stsc2013/tech-12E.pdf on February 28, 2013.
14. UNCOPUOS, "Long-Term Sustainability of Outer Space Activities," accessed at http://www.unoosa.org/pdf/ limited/c1/AC105_C1_2010_CRP03E.pdf, on February 28, 2013.
15. UNOOSA, "National Space Law Database," accessed at http://www.unoosa.org/oosa/en/SpaceLaw/national/ state-index.html on March 3, 2013.
16. Minister of State's Reply in Lok Sabha on February 23, 2013, to Unstarred Question No. 369 on Indian activities in space, accessed at http://www.isro.org/parliament/2011/Budget/LUSQ369.pdf on March 6, 2013.
17. "India and Space Laws: A Millennium Perspective," March 9, 2000, *The Hindu*, accessed athttp://www.hindu. com/2000/03/09/stories/08090008.htm on March 7, 2013.
18. Dr. Ranjana Kaul, in a presentation titled "National Space Legislation: A Blueprint for India," at the 3rd Regional Space Conference on Bringing Space Benefits to the Asia-Pacific Region, Bangalore, June 27-29, 2005.
19. Brian Harvey, Henk H. F. Smid and Theo Pirard, *Emerging Space Powers: The New Space Programs of Asia, the Middle East, and South America* (Chichester, UK: Praxis Publishing Ltd., 2010), p. 250.
20. "Overview of Indian Space Sector 2010," accessed at http://mycii.in/KmResourceApplication/E000000056. 662.Antrix-CII-Deloitte%20report%20on%20Indian%20Space%20Sector%202010.25Aug10.pdf on October 19, 2012.
21. Kaul, n. 18.
22. http://dos.gov.in/space-policy.aspx.
23. ISRO, "List of Foreign Satellites Launched by PSLV," accessed at http://www.isro.org/pdf/foreignsatellite.pdf on March 7, 2013.
24. Ranjana Kaul and Ram S. Jakhu, "Regulation of Space Activities in India," in Ram S. Jahku, ed., *National Regulation of Space Activities* (London: Springer, 2010), Ch. 8, p. 194.
25. Ibid., Ch. 8, p. 166.
26. Ram S. Jakhu and Yaw Otu M. Nyampong, "International Regulation of Emerging Modes of Space Transportation," in Jakhu, ed., n. 24, Ch. 17, p. 223.
27. Michael Sheehan, *The International Politics of Space* (Oxon: Routledge, 2007), Ch. 8, pp. 129-135.

28. Cesar Jaramillo, ed., *Space Security Index 2012* (Ontario: University of Waterloo, 2012), p. 18, accessed at http:// www.spacesecurity.org/SpaceSecurityReport2012.pdf on March 20, 2013.
29. "UCS Satellite Database", accessed at http://www.ucsusa.org/nuclear_weapons_and_global_security/space_ weapons/technical_issues/ucs-satellite-database.html on July 26, 2013. The report also lists 523 in LEO, 75 in MEO and 38 in polar orbits, totalling 1071 operational satellites.
30. Gbenga Oduntan, *Sovereignty and Jurisdiction in the Airspace and Outer Space* (New York: Routledge, 2012), Ch 8, pp. 211-212.
31. ITU, Radio Regulations Article S23.13.
32. Ram S. Jakhu, "Legal Issues of Satellite Telecommunications, the Geostationary Orbit, and Space Debris," *Astropolitics,* 2007, p. 182.
33. Sheehan, n. 27, Ch. 8, pp. 135-138.
34. M. Y. S. Prasad, "Space-Based Telecommunications including Tele-Education & Telemedicine- Implications to the Area of Space Law," in Proceedings of the Space Law Conference 2005: Bringing Space Benefits to the Asian Region, Bangalore, India, June 26-29, 2005.
35. Michael Listner, "Legal Issues Surrounding Space Debris Remediation," *The Space Review*, August 6, 2012, accessed at http://www.thespacereview.com/article/2130/1 on April 17, 2013.
36. James Clay Moltz, "The Code of Conduct: A Useful First Step," in Ajey Lele, ed., *Decoding the International Code of Conduct for Outer Space Activities* (New Delhi: Pentagon Security International, 2012), Ch. 27, p. 144.
37. Joan Johnson-Freese, "Supporting an International Code of Conduct," in Lele, ed, Ibid., Ch. 26, pp. 141-2.
38. Li Juqian, "The Future of the International Code of Conduct for Outer Space Activities: From Mission Impossible to Mission Sustainable?," in Lele, ed., Ibid., Ch. 21, p. 120.
39. Moltz, n. 36, Ch. 27, p. 144.
40. Jana Robinson, "Europe's Space Diplomacy Initiative: The International Code of Conduct," in Lele, ed, n. 36,, Ch. 4, p. 29.
41. Juqian, n. 38, Ch. 21, p. 120.
42. Hillary Rodham Clinton, US Secretary of State, INTERNATIONAL CODE OF CONDUCT FOR OUTER SPACE ACTIVITIES, Press Statement, Washington, DC, January 17, 2012.
43. Michael Listner, "Code of Conduct: Corrections, Updates, and Thoughts Going Forward," June 18, 2012, accessed at http://www.thespacereview.com/article/2101/1 on April 26, 2013.
44. Michael Listner, "EU Code of Conduct: Commentary on Indian Concerns and

their Effects," November 28, 2011, accessed at http://www.thespacereview.com/article/1977/1 on April 26, 2013.

45. Tiffany Chow, "The Space Code of Conduct, and Other Factors," September 3, 2012, accessed at http://www.orfonline.org/cms/sites/orfonline/modules/analysis/AnalysisDetail.html?cmaid=41445&mmacmaid=41446 on May 1, 2013.
46. Listner, n. 43.

7

National Security from the Space Domain: Legality and Ethics

Diplomacy without arms is like music without instruments.

— Frederick the Great

One hundred hours! That was all it took for the ground offensive to achieve its objectives in the 'mother of all battles'. The ground offensive in Operation Desert Storm, the first Gulf War to liberate Kuwait from Iraq commenced on February 24, 1991, after five weeks of intense aerial bombardment. The operation led by the US and UN-authorised coalition force of over 30 countries wound up within 100 hours of the ground offensive having liberated Kuwait and advanced into Iraq. Space support in no way was the centrality of the coalition victory but this was the first time in history that space power transformed warfare and played such a prominent role in support of the air-land battle that prompted one US General to comment, "Desert Storm was a watershed event for space systems...space owned the battlefield...we had the inherent spacecraft flexibility to alter our operations to support specific needs of the terrestrial war-fighter"[1] while another General viewed the war as the "the first space war"[2] even though no war took place in space. This was the war where military activities in space moved from the realm of 'force enabler' to that of 'force multiplier'. As many as 60 military and commercial satellites were pressed into service for the war effort, which supported the coalition in almost the complete spectrum of military-support activities of reconnaissance, intelligence gathering, weather monitoring, missile launch warning, weapon guidance, search and rescue, imaging and GPS. Nearly 90 per cent of inter-theatre and intra-theatre secure communications were provided by satellites.[3] Space now had been firmly established worldwide as the fourth dimension of war.

That the military has a very large presence in space was never in doubt. Space has always been militarised as military considerations were the roots on which space ventures were founded. Global space budgets have been level at US$ 70 billion annually for the past decade, of which US$ 30 billion is for military purposes.[4] Some satellites do not conform to the conventional classification of either military or civil use. As with satellites, space technologies and hardware too operate in the undefined or dual-defined area between the military and civil – 95 per cent of space technologies are dual use in nature[5] seeming to suggest that militarisation of space is a *fait accompli* – no longer an issue for debate but an unverifiable reality which cannot be wished away.

This chapter will trace out the genesis of military activities in space that had its roots in the Cold War rivalry between the two superpowers and led up to the full spectrum of activities that contribute to a nation's security apparatus. The legal issues that govern militarisation of space will be profiled, including activities that are outrightly banned or 'conveniently' permitted, under the contemporary legal regime.

A Historical Perspective

As man began his quest for conquering land, he took to capturing the *high ground* for the inherent tactical advantages that accrued. Not only was the acquisition of the high ground a necessity for dominating the area around it, but it also gave the conqueror a higher platform for surveillance. This high ground gave him the capability of observation, providing him early warning of the threats that could manifest against him. As the art of war-fighting evolved, he built his defences on this ground of tactical significance, and to enable early warning, he extended his line of sight or range of observation by raising the platforms of observation and positioning them beyond his defences. It was, thus, only a matter of time before man would look to the skies and seek to place a military platform in the air.

The two world wars brought in the role of air power, both in terms of raising the high ground and increasingly employing air power in support of the ground war. The bar was further raised towards the end of World War II when Germany heralded mankind's first step into the space age by using V-2 rockets. Post World War II, the future of war-fighting was based essentially on two products of the war: the nuclear weapon and the means of delivering it. The US had already conclusively demonstrated and established its lead in the former and towards the latter, had long range bombers to deliver them unlike the Soviets. The Soviets turned to ballistic missiles as an alternative means of delivery. *The genesis of the thought of venturing into space had always been military oriented – an offshoot of the arms race – and commenced with ballistic missiles.* The arms race led to space which then led to the space race. Developments in powerful rockets led to leveraging the technology of ballistic missiles to place a satellite in orbit. The Soviet Union launched the world's first satellite in space, the Sputnik I, in 1957. The space race had truly begun and the first American satellite was launched four months later. Space launch technology stood proven and national confidence restored – the next step was launching military satellites.

The Birth of Military Satellites

The era of the Cold War was characterised by intense competition and mutual suspicion, and the new high ground in space offered a very viable vantage point as it offered tremendous potential in watching over missile installations, deployments and activities of the other. The US' need of a strategic reconnaissance satellite to watch over the Soviet Union and other potential adversaries arose due to a combination of events that commenced with being caught by surprise by the Pearl Harbour attack, the Soviet Union's developing nuclear bombs and North Korea's surprising invasion of South Korea in 1950.[6]

The US created the world's 'first' in military satellites (milsats). The first experimental reconnaissance satellite or spy satellite, Corona,[7] codenamed Discoverer 1, was launched by the US in February 1959. The spectrum of applications in support of the military was enlarged and the Discoverer was followed by the US launching the world's first military navigation satellite, Transit 1B; the first weather satellite, TIROS-1, and the first military communications satellite, the Echo-1, all in 1960. The Soviets followed with their first military photo-reconnaissance satellite, the Zenit, in 1961. The era commencing the 1960s, witnessed satellites being used as the first credible source for arms control verification. The decade saw frenzied launch activity by both the superpowers as both aimed at fielding technology – and capability – enhancing spacecraft with better resolution and more accuracy that could contribute to conventional force enhancements. Commencing in the 1960s, both the nations gravitated towards deploying space-based missile-launch early warning systems.

The GPS, so widely proliferated as a global civilian and commercial utility in today's life, actually owes its origins, just like the Internet, to the military. The world's first military navigation system satellite, the Transit, launched by the US in 1960, enabled positional fixing and locating ballistic missile submarines to within 200 m accuracy. The system was opened for access by civilian users in 1967.[8] Technological developments thereafter continued to increase the military's exploitation of space. Two new fields in space that were added to further support the military were the fields of meteorology and communications. All such space-based military-support systems drastically enhance the quality and quantity of data for planning and execution of military missions on Earth, and battlefield awareness to the military commanders.

With the collapse of the Soviet Union in 1991, the space race as good as ended but the US' thrust on building its space capabilities in support of its military continued. Two reasons can be ascribed to this phenomenon. Traditional rivalry with the Soviet Union no longer being the motivating validity for its pursuits in space, the US increasingly began to look at itself as a global watchdog, with geopolitical interests that transcended its Soviet-centric outlook – an outlook that had been the cornerstone of policy-making hithcrto. Capability-building in space, thus, was essential to the US' expanding global dominance and influence. The second factor that loomed on the horizon was China's emergence in the space arena – first as a regional power and then, gradually, backed by a rapidly growing economy, as a competitor and a challenger to the US' hegemony in space.

The US' 'pole position' in space support infrastructure can be gauged by data provided by the Union of Concerned Scientists. As of May 31, 2013, the US has 459 satellites out of a total of 1,071 operational satellites while Russia and China trail at 110 and 105 respectively. Investments made by the US in its military space programme are indicated by the fact that 131 of the 459 American satellites are milsats[9] duly supported by a military space programme budget that is considerably larger than those of all other spacefaring nations combined and about 20 times that of Russia.[10] The US today is not only the most powerful in space but also the most dependent on it.

Following the lead of the two traditional rivals and in response to their own national security concerns, countries like China, France, the UK, Germany, Japan and Israel currently operate milsats and dual-use satellites.

The Growing Reliance on Milsats in Conflicts Worldwide

By the early 1960s, it was established that reconnaissance satellites were the only non-intrusive source for gathering intelligence, seemingly conducting their operations from the safety accorded by the altitudes of LEO. Initial operations of such spy satellites hinged upon detection of the other's deployment and move of nuclear delivery means, viz., ICBMs, nuclear submarines and strategic bombers. It was the Cuban missile crisis in 1962 that highlighted their role in actual conflict. The US' Corona satellite imagery revealed reasonable estimates of Soviet nuclear tipped artillery rockets and combat aircraft in Cuba while the Zenit provided to its Soviet commanders an estimate of the US naval might off the coast of Cuba. The fear of a nuclear showdown led to a detente, flagging the role that satellites played in nuclear-dissuasion. American military commanders in the Vietnam War were provided imagery from the Nimbus reconnaissance satellites. A daring raid in 1970 to rescue American prisoners at Son Tay was made to coincide with a break between two tropical storms over Vietnam – a mission conceived based on weather data provided by meteorological satellites of the US Defence Meteorological Satellite Programme (DMSP), showcasing the switch in the role played by milsats from the strategic to the operational.

It was the Soviets who brought satellites closer to supporting the land war by employing them in a tactical role. In the Yom Kippur War of 1973, the Soviets launched two Cosmos reconnaissance satellites that kept the Egyptians abreast of Israeli breaches on the Suez Canal and five satellites more after the war to monitor the ceasefire agreement.[11] This was the war in which the US had denied Israel's request for satellite imagery, and post-war, led to Israel's resolve to indigenise its space efforts. Improved relations between the two thereafter enabled sharing of satellite imagery with Israel for the conduct of their Entebbe operation in 1976 and the bombing of Iraq's Osirak nuclear reactor in 1981.[12]

The US' support to Britain in the Falklands War of 1982 against Argentina on the other hand, was more forthcoming. The US not only launched new satellites but also manoeuvred currently orbiting ones to provide near real-time imagery of the Argentinean troop and naval movements, and mapping likely landing sites for the British offensive.

Satellite based force-multiplier enhancements continued to be fielded with ever increasing effect in the first Gulf War (1990), Operation Allied Force against Yugoslavia's ethnic cleansing (1999) and Operation Enduring Freedom against the Taliban regime in Afghanistan (2001 onwards) under the umbrella of the Global War on Terror. It was, however, 2003 onwards in Operation Iraqi Freedom, with the aim to oust Saddam Hussein, that the US' economic and technological dominance in space stood proven. It was not just the 100 odd satellites that contributed to the war effort, but the manner in which they were integrated to achieve the objectives that was the hallmark of their employment. The US' Brig. Gen. Larry James, Senior Space Officer, commented on the role that space played, "It's a combined and joint integrated operation, with space supporting all the other forces and activities."[13] For the first time, data was received from a multitude of sensors and a variety of platforms, and fused together to provide a cogent picture to the tactical commander. Besides the fields of communications, weather, surveillance, mapping and missile warning, the enhancements in the field of targeting stood out starkly. The sensor-to-shooter time of two hours as obtained in Operation Desert Storm a decade earlier was brought down to 11 minutes in this war[14] enabling real-time satellite feeds into the cockpit.[15] Use of precision guided munitions rose from 8 per cent to 68 per cent and data handling speeds grew 10 times in the span of a decade.[16] On the flip side, the reliance on commercial satellites too grew with about 60 percent of the communications and data transmitted by the allied forces being routed through commercial satellites.[17] The ever increasing reliance of US conventional forces on space assets led the Hon. Peter B. Teets, Under Secretary of the US Air Force, to remark, "Space systems are inextricably interwoven into the fabric of America's national security."[18]

Militarisation vs Weaponisation of Space

A satellite has no conscience.

— Edward R. Murrow
American TV News broadcaster

Militarisation of space in simplistic terms is the employment of space-based support systems as described briefly above, which are used by the military in support of its land, sea and air operations. In that sense, space has been militarised ever since the first milsat took to the skies and shall remain so in the foreseeable future. Weaponisation of space, on the other hand, is victim to unagreeable definition but generally understood as placing of space assets with a destructive capability that can be aimed against other assets in space or on the Earth.

It is reasonable to state that space has not been weaponised so far, insofar as use of weapons in orbit is concerned. As satellites continued to be launched during the Cold War, it was natural to assume the worst case that, in times of a war in the three dimensions, the fourth would invariably get involved. The thought of space, not just supporting the war effort but becoming a battlefield by itself, was fearsome enough, necessitating the two superpowers to look at satellites as assets that needed safeguarding, as well as being used as weapons. Satellites as offensive weapons were designed, tested, mothballed and may be even launched[19] but never used in space. Various types of 'killer' satellites and platforms, and the technologies that contribute to their effectiveness, have already been covered in detail in Chapter 5.

Though not qualifying as weapons in space, ground- and air-launched missiles have been launched into space in the past for testing of ASAT technology. Both the US and the Soviet Union experimented with ground- and air-launched weapons to blind or destroy each other's space assets, tests that continued till the mid-1980s when both the powers applied a self-imposed moratorium on further testing due to the threat posed by space debris that resulted from the tests. The only other country that has demonstrated its ASAT capability is China when it launched a ground-based missile interceptor to destroy its weather satellite, the 750 kg FY-1C in LEO, in 2007.

Milsats in the Realm of National Security: Some Emerging Trends

Space is being viewed as the military centre of gravity, at least by the select few space-capable nations and this seminal thought defines the employment of space as a force multiplier which multiplies the effectiveness of combat forces in the terrestrial domain. Tremendous advancements in space technology have enabled the US to rely on space for 'full spectrum dominance' of its land wars while other prominent space players achieved their goals of space-support by way of varying priorities and to varying degrees.

In all the cases, space-support systems have enhanced the battle awareness among the terrestrial commanders and established the need for 'full spectrum' support in real-time communications, navigation and precision guidance, reconnaissance and intelligence gathering, weather monitoring and vitally, in the field of early warning systems – systems that monitor an adversary's launch of a ballistic missile and conduct of activities towards nuclear detonation. Such systems exploit the complete orbital spectrum that technology enables mankind to do, which manifests itself in two consequential fallouts. The first is the economics of the space launch industry, where heavy investment in the military-industrial complex has resultantly been able to reduce the cost of an individual launch which has a direct impact on the commercial launch sector. The second significant result of military-support activities is the enormous amount of 'spinoff' generated by technologies, both in the field of civil-scope space activities and in the realm of non-space commercial sector.

While the US enjoys the status of the sole superpower with an assailable lead in 'full spectrum dominance' in space in support of its 'network-centric warfare' on the Earth, the narrative in respect of Russia in the post-Cold War era has been unsurprisingly different. Russia inherited the world's second largest fleet of satellites which by the turn of the century had between 70 and 80 per cent satellites nearing their designed lifespan.[20] Economic compulsions necessitated prioritisation. Russia focussed first on its early warning systems and then on the GLONASS navigation systems. The past decade has seen Russia stressing on maintaining and protecting its space assets. It was only in 2006, the first year of the 10-year federal space programme that saw the Russian military space budget increase by as much as one-third,[21] not quite adequate enough to revamp most parts of its military space ventures. Russia has lost five communications satellites and one Progress cargo spacecraft since December 2010, leading to the Russian Parliament's Audit Chamber in July 2013 criticising the *Roscomos* space agency for its inefficiency.

Milsats and Orbital Employment

Various space based services that are provided for the military from various orbits are listed as under (Table 1.7):[22]

Table 1.7

Orbit	Altitude (Kms)	Comms	Recon/ Imaging	Navig	Meteorology	Radio Intelligence	Missile Warning
LEO	Up to 2,000	✓	✓	✓	✓	✓	
MEO	20,200	✓	✓	✓			
GEO	35,800	✓				✓	✓

The data above indicates that military-support services employ all the orbits currently engaged in civilian services while military-specific services like missile-launch and nuclear detonation warning systems appear unique to the GEO. The field of communications is omnipresent – employed in all the three orbits.

Communications: Support from Milsats

Military satellite communications systems comprise the single most important military space capability[23] in the contemporary context. Not only do such milsats enable military commanders to exercise command and control over their forces deployed worldwide by being networked to provide real-time information, but also are of immense use in transfer of huge amounts of data to and from sensors and control centres over encrypted communications links. Communications and data-handling are key ingredients in a military's network-centric warfare. Amongst the notable milsats currently operational, the US' Military Satellite Strategic and Tactical Relay (Milstar) System is based on a constellation of five cross-linked milsats in GEO which provides secure communications and data relay functions to the three Services of the US military. This system is under replacement by a constellation of six Advanced Extremely High Frequency (AEHF) satellites, two of which have already been placed in GEO. Each of the AEHF satellites has greater total capacity than the entire five Milstar satellites. The Defence Satellite Communications System (DSCS), the workhorse of the US military's super-high frequency communications, is a hardened and jam resistant constellation based on nine satellites in GEO that handles high-priority command and control messages to battlefield commanders. To meet the growing demands for bandwidth, the DSCS is planned to be replaced with the Wideband Global Satellite System (WGS) based on ten satellites which, compared to DSCS, will provide 10 times the data handling capacity in the X and Ka bands.[24] The fifth satellite in the series was launched in May 2013.

The field of military communications has increasingly been placing reliance on the commercial space sector to meet its burgeoning demand for bandwidth, as evidenced in Operation Iraqi Freedom where 80 per cent of the military's communications need was leased from commercial sources. With the aim of saving an estimated US$ 800 million[25] that would be spent annually to lease capacity on multiple commercial satellites, a new initiative by the US DoD plans to lease a single communications satellite operating in the Ka and Ku bands for 15 years. The proposal is currently facing problems of the requisite funding.

Russia maintains several communications systems, most of which are dual-use. One such system is the Raduga constellation which also provides secure military communications channels. The Geizer constellation planned on four communications satellites in GEO acts as the relay system for remote sensing and communications satellites and is being replaced by the new-generation Garpun series of relay satellites, the first of which, the milsat Cosmos 2473, was launched in 2011[26] tasked specifically to relay imagery data from reconnaissance satellites in LEO to Russian ground stations. In addition, the Rodnik series of milsats operate as 'store-and-dump' communications satellites operating in LEO.

China's military satellite communications are believed to be provided by the DFH series of satellites. ChinaSat 22 launched in 2006 and catalogued as a communications satellite, is believed to be a military satellite named Feng Huo 2, a part of the data-relay network system that enables theatre commanders to communicate and share data with other forces under joint command. France, a leader with the most advanced and diversified independent space capabilities in the EU space community, operates the milsat Syracus system and the dual-use Telecomm satellites. The UK has provided for its military a secure high-bandwidth capability through 2022 by the Skynet-5 satellite, while Spain's Spainsat is a dedicated milsat providing X- and Ka-bands services to its forces. Japan's military communications in the Ku- and Ka-bands are met by the commercial Superbird satellites in GEO while South Korea relies on its dual-use Koreasat-5 (Mugunghwa 5) satellite that provides the Ku-, C- and military SHF bands.[27]

The Indian communications satellite, the GSAT 7, was launched into GEO on August 30, 2013, aboard the Arianne and is the first to be publicly championed as a pure military satellite – one which is dedicated for the communications needs of the Navy. This is to be followed by similar satellites for the Army and the Air Force.

Milsats for Intelligence, Surveillance and Reconnaissance

The very first of the world's milsat was the Corona in 1960, the US' photo-reconnaissance satellite armed with a camera whose film had to be recovered from orbit. Twelve years of the Corona's programme helped to view and photograph the 'denied' landmass of the Soviet Union and more significantly, provided intelligence that laid the foundation in verifying disarmament agreements. Surveillance satellites differ marginally in the context of their definition. Operating in the realm of early-warning satellites, they monitor parts of the Earth for hostile military activities of the likes of missile-launch and nuclear detonations.[28] Functions performed by the three kinds of satellites are generically clubbed as Intelligence, Surveillance and Reconnaissance (ISR) support to the military.

Over time, the resolution of the cameras aboard reconnaissance satellites improved to sub-metre and by the mid-1970s, digital imagery was being beamed back live. Current resolution capabilities of operational milsats, of course, remain classified but the US Improved Crystal satellites are believed to have a resolution of six inches[29] while the European Helios milsat provides a resolution of 30 cm,[30] accurate enough and ideally suited for targeting purposes in the tactical domain. Milsats today that provide ISR functions to terrestrial commanders include imagery, Signals Intelligence (SIGINT), Electronic Intelligence (ELINT) and multi-spectral data in the region of IR and UV bands. Radar satellites equipped with Synthetic Aperture Radar (SAR) detect forces and activities on the ground in all weather conditions, both during day and night.

The need for Early Warning (EW) satellites arose from the requirement of prompt warning of an ICBM launch during the Cold War which might necessitate a strategic response. One of the drivers for EW satellites technology capitalised on detecting the characteristic IR signature of an ICBM's exhaust plume in the boost phase in a typical flight trajectory lasting about 30 minutes. Availability of adequate warning time during the Cold War, as part of Ballistic Missile Defence (BMD), allowed the two superpowers to move away from 'hair trigger' alert postures. IR detection technology even differentiates between solid and liquid-fuelled missiles[31] and has evolved to discern launch of even short range missiles. The US' Defence Support Programme (DSP) fulfils the role of EW through its satellites in GEO and played an active role in warning of Iraq's Scud attacks in the Gulf War. Early attempts at nuclear-detonation warning had payloads that carried electromagnetic pulse detectors in addition to IR detectors onboard the US' Vela satellites as part of the DSP. The DSP is in the process of being replaced by next generation Space-Based IR System (SBIRS), the first satellite of which was launched in 2011. Conceived around four satellites in GEO and four payloads on satellites in highly-elliptical orbits, the IR sensors are three times more sensitive than those in DSP and aim to reduce warning time to 10-20 seconds as against 40-50 seconds obtained through DSP.[32] A feature of this system is that while large swaths of land will be scanned for missile launches by the onboard scanning sensor, the staring sensor will focus on specified areas of interest with superior sensitivity.[33] In addition to missile defence, SBIRS will support missions such as Space Situational Awareness (SSA) to forces on the ground.

By the turn of the century, Russia maintained four Oko early warning satellites in High Earth Orbit (HEO), covering the US' land-based missile launch fields, a capability adequate to detect massive attacks but unreliable against individual missile launches.[34] This is complemented by a next-generation EW satellite, the Prognoz, in GEO, capable of detecting missiles against the background of the Earth.[35] In the field of reconnaissance, Russia launched its most advanced milsat, the Persona 2 in June 2013 that provides electro-optic imagery of 0.5 m resolution, and in the same month, the Kondor satellite with a SAR payload that provides a resolution of 1-3 m.

Countries that do not operate dedicated milsats, use civilian and commercial space assets for the spectrum of force enhancement for their national security. Such non-military satellites currently have the technological capabilities that enable them to gainfully contribute and play a significant role in providing ISR data, the most prolific of military-support ventures, to its military and, thus, are classified as dual-use satellites.

Table 7.2 below quantifies the minimum resolution standards required to support various military force enhancing ISR activities:

Table 7.2: Minimum Resolution Required (in metres) for Surface Objects[36]

Target on Ground	Detection	General Identification	Precise Identification	Technical Analysis (cm)
Vehicles	1.5	0.6	0.3	4.5
Aircraft	4.5	1.5	1.0	4.5
Nuclear Weapon Components	2.5	1.5	0.3	1.5
Rockets & Artillery	1.0	0.6	0.15	4.5
Command & Control HQs	3.0	1.5	1.0	9.0

In China's context, satellites that have the requisite resolution and are likely being used by its military are: the Zi Yuan series of Earth observation satellites which could support tactical reconnaissance and surveillance, the Tsinghua microsatellites with multi-spectral camera and a high-resolution panchromatic imager and the Yaogan series of six satellites, two of which carry the all weather, day/night SAR. The technological capability of its reconnaissance assets is proven by the fact that the world's first anti-ship ballistic missile, the 1,500 km Dong Feng-21, relies on reconnaissance satellites for targeting and tracking.[37]

France has the lead among the EU members in the field of milsats that provide ISR intelligence. The Essaim constellation has four SIGINT satellites in orbit since 2004 and has followed it up with four Elisa demonstration microsats launched simultaneously in 2011, which monitor transmissions of ground-based civil and military radars. The operational component of this system is planned to be launched by the end of this decade. A resolution of 30 cm is provided by the pair of Helios milsats in the field of image intelligence.

Driven by common concerns about Europe's security and budget restrictions, defence cooperation amongst EU members has been a key driver that led to independent space assets being shared and networked together. Germany's first milsat, the Sar-Lupe, with a SAR payload, along with Italy's COSMO-SkyMed radar satellites is being integrated with France's Pleiadus dual-use optical remote sensing satellite network that provides a resolution of 50 cm,[38] enhancing the network to now provide multi-spectral support capability. The Pleiadus satellite itself was a joint venture with Austria, Belgium, Spain and Sweden, partly funded to gain access to its imagery.

Driven by its unique geostrategic security compulsions, Israel has a fairly extensive multi-spectral spread of dedicated milsats. The Ofeq-9 series of reconnaissance milsats, all launched indigenously aboard the Shavit launcher, provide a 50 cm resolution (through Ofeq-9) which is further augmented by two dual-use Eros imaging satellites. In SAR capabilities, Israel operates one of the world's most advanced reconnaissance milsat, the TecSAR, launched aboard the Indian PSLV in 2008, which provides a resolution of 10 cm.[39]

Japan's military endeavours have originated rather hesitatingly, prodded by the belligerent activities of North Korea when it launched a ballistic missile in August 1998 on a trajectory that overflew Japan and then galvanised by a rising superpower in the neighbourhood, China, finally prompted Japan's legislature, the Diet in 2008, for the first time, to allow its military the use of space[40] which hitherto was a purely civil-oriented programme. The ISR requirements of the military are met by the dual-use "information gathering" series of remote sensing satellites, two of which are optical imaging and the other two radar satellites, that provide one metre resolution imagery to its military.[41]

Quite akin to Japan, India is a country whose space ambitions were driven by socio-economic developmental needs and later had geopolitical considerations impacting its national security which led to modifying its space launchers to build ballistic missiles. It has no *stated* objectives of space-support for its military but has space assets which provide the kind of high resolution imagery that could be force enabling, if it so chooses to. India operates the Cartosat series of Earth observation satellites, of which the highly agile and steerable Cartosat 2B launched in July 2010 provides a resolution better than one metre.[42] The SAR capability is played out in two radar imaging satellites (RISATs 1 and 2) which provide all weather day and night radar imaging capability with a resolution of 3 m.

Navigation and Precision Guidance from Space

These satellites provide positional information that is integrated into all facets of military operations, from indirect strategic support to military forces deployed worldwide to tactical support in real-time in target identification, precision weapons' guidance, and search and rescue missions. The US' NAVSTAR GPS, operational since 1993, uses a constellation of 24 to 32 satellites at 20,200 km altitude, communicating with control stations and users on the L band. Satellites in this constellation comprise the fourth generation Block II satellites while Block III satellites with a design life of 15 years planned for launch commencing 2014 will transmit at higher power levels to reduce vulnerabilities to jamming.

Three-dimensional information of position (longitude, latitude and altitude), time and velocity is provided to military users by at least four satellites 'in sight' though redundancy and accuracy are further enhanced by additional satellites. Communications to military users undergo encrypted encoding while those for commercial entities follow a public encoding that enables lower resolution navigation. Positional accuracy of a few metres has had a significant impact in the field of munitions' delivery and employed in ICBMs, cruise missiles, ground-to-air aircraft and artillery munitions, enables milsats to play a significant role at the tactical level of warfare. GPS as add-on kits attached to air-delivered 'dumb' bombs enables pin-point accuracy that prevents collateral damage and conserves munitions' wastage rates.

Soviet efforts at the GLONASS satellite navigation system began in 1982 which was fully operational with a global coverage in 2011 with 24 satellites at 19,900 km orbit and is the only other system after GPS that currently provides a global coverage. Post the break-up of the USSR, the programme was in limbo till Russia recognised this area of space activity as one that required prioritisation in budget and effort. US$ 4.7 billion in the period 2001-10 allocated by Russia into this venture rejuvenated the programme. The satellites' designs have undergone several upgrades, with the latest version being the third-generation GLONASS-K with a life of 10 years.

The GPS and GLONASS were one of the byproducts of the space race during the Cold War rivalry between the two superpowers. Fearing GPS services being denied access by the US as a fallout of political disagreements or in times of conflict, other countries that ventured into the field of space-based navigation systems thereafter, did so to retain their independence and deny the vulnerability that such reliance would have resulted in.

China operates the Beidou regional positioning satellite system, with satellites in a mix of MEO and GEO orbits, which provide 10 m accuracy for China and its immediate neighbourhood.[43] This navigation system then will graduate onto the Compass Navigation Satellite System which aims for global coverage with 35 satellites by the year 2020[44] of which 14 are already operational as of December 2012. With positional accuracy in decimetres[45] the system has tremendous potential in the realm of the military.

Europe's initiative in this field is the Galileo system. A joint initiative of the EU and the ESA, the US$ 6.45 billion programme envisages a constellation of 30 satellites in MEO providing accuracy within one metre[46] aiming for completion by 2019. The first two of its satellites were launched in 2011 and two more added a year later. Incidentally, China had joined the Galileo project in 2003 but opted out three years later to independently pursue its Beidou/Compass system. There is an ongoing conflict in the use of frequencies employed by the Chinese and the European navigation systems due to an overlap in the two spectrums. Neither of the two is in violation of international regulations as no interference has resulted, but the implication is that neither of the two can jam the other's system in times of conflict without compromising the continuity of its own.[47] With respect to compatibility with GPS, Block III satellites of the US' GPS will contain standardised features that will enable the Galileo and GPS to inter-operate, providing enhanced precision to users.

India's stance on its self-reliance has been echoed in its indigenous navigation satellite programme, the Indian Regional Navigational Satellite System (IRNSS) being fielded by ISRO. Approved in 2006, the project envisages a constellation of seven navigation satellites, three in geostationary and four in geosynchronous orbits, beaming navigation signals in the S band (2-4 GHz) and ranging signals in the C band, providing 10 m accuracy over the mainland.[48] The first satellite of the IRNSS constellation, the IRNSS 1, was launched in July 2013 aboard the Indian space-horse, the PSLV, while the full constellation is planned to be operational in 2014.[49] It will offer two services: Standard Positioning Service and Restricted Service with encryption. The system will be compatible and inter-operable with the Galileo and GPS.[50]

Japan, on the other hand, has followed a slightly different path. It has commenced developing an indigenous regional navigation programme, albeit one that supplements the GPS rather than substituting it. GPS signals are only available about 90 percent of the time in Japan, but satellite navigation will be possible 99.8 percent of the time with the Quasi Zenith Satellite System (QZSS) underway.[51] Japan orbited a QZSS satellite, the Mitchibiki, in 2010. As Japan's geographic latitudinal position does not accord the benefits accruing out of placing a satellite in geostationary orbit above the country, the QZSS, at an inclined orbital plane of about 40 degrees,[52] banks upon a number of such satellites at near-zenith over Japan, providing a sub-metre positional accuracy. US$ 526 million has been contracted for construction of three more QZSS satellites that will complete the constellation by 2017. The system not only enjoys tremendous commercial usage by increased reliability and accuracy complementary to the GPS, but is dynamically scalable to provide military communications by equipping them with communications and broadcast sub-systems.[53]

Meteorology Satellites

The civilian service of understanding and predicting weather from space found tremendous potential for the military. Global capability to monitor clouds and weather patterns augmented the military's planning and execution of combat and reconnaissance missions, and of sea-surface conditions in support of naval operations. The most extensive weather capability is exhibited by the US' meteorological satellites programme spanning a period of almost five decades. The Defence Meteorological Satellite Programme (DMSP) commenced in the mid-1960s but its original military motive was declassified and revealed only in 1973 and placed under the control of the Department of Commerce in 1998, thus, merging the civil and military weather programmes. The system operates its Block 5 satellites at Sun-synchronous polar orbits at an altitude of 830 km which provide continuous visual and IR imagery of cloud cover and carry sensors with improved capabilities. The life of these satellites was designed to last till 2010 and currently the US is faced with gaps in weather monitoring due to delays in launching replacements and inadequate funding.[54] The civil and military components of space-based weather systems have now been divorced under the Obama Administration and the planned replacement for the military was to be the Defence Weather Satellite System (DWSS), at US$ 5 billion through 2015, with the first launch slated for 2018.[55] The DWSS was cancelled in 2012 for lack of funds leaving DoD officials deliberating on the appropriate future path, while the gap in weather data provisioning now will need to be filled up in the interim with launching the balance two legacy DMSP satellites, DMSP-19 and -20, to provide weather services into the mid-2020s.[56]

The European initiative in this field is the European Commission's Sentinel satellite programme that will launch six families of Sentinel satellites that include SAR and multi-spectral payloads, commencing late-2013 through 2020, as part of its Global Monitoring and Environmental Security (GMES) programme. The new Sentinel series satellites will augment Earth sensing services provided by current satellites like the French Spot to support land and maritime surveillance missions, and border control.[57]

Space and National Security: The Legal Dimensions

International treaties and agreements that make-up internationally accepted space laws have been adequately covered in the previous chapter. This section aims to highlight those aspects of the accepted law that deal specifically with either enabling or constraining the conduct of space activities that contribute to the militarisation of space. Viewing the issue in the context of weaponisation of space, it would be prudent to briefly look at the issue of weaponisation and the mechanisms that govern the use of the weapons, insofar as they impinge upon employability in space. Some of the treaties like the Anti-Ballistic Missile (ABM) Treaty, Strategic Arms Limiting Talks (SALT) and Strategic Arms Reduction Treaty (START) are bilateral instruments between the two superpowers driven by the need to limit strategic weapons armed with nuclear payloads. Seen as vestiges of the Cold War era, they are no longer of any significant relevance but for their seminal influence in limiting weapons in space and setting the tone for other multilateral mechanisms that followed.

The Limited Test Ban Treaty (LTBT), 1963

One of the foremost arms control treaties, the LTBT was signed in 1963 to slow down the arms race and prevent contamination of the Earth's atmosphere. It prohibits nuclear detonations, which include both weapon tests and tests for peaceful purposes, in the atmosphere, in outer space and under water. It does not outrightly ban underground testing as long as the fallout is contained within the territorial limits of the state conducting the test. The LTBT is also generally referred to as the Partial Test Ban Treaty (PTBT) and the Nuclear Test Ban Treaty (NTPT). The treaty as of February 2013 has 126 member states with France, China and North Korea being prominent non-signatories. This is the treaty that introduced banning of nuclear explosions in outer space *after* both the superpowers had already conducted nuclear tests in space but *before* the base legislation for activities in outer space, the Outer Space Treaty, was brought into vogue in 1967. The follow-up instrument to the LTBT, the Comprehensive Test Ban Treaty (CTBT) adopted in 1996 by the UN, bans nuclear explosions in all environments.

The Outer Space Treaty (OST), 1967

Adopted in 1967 by the UN, OST was the first treaty that enunciated regulations exclusively for the space domain. The centrality of this base regulation lies in establishing outer space as a medium common for all mankind for exploitation for peaceful purposes. Some facets of this treaty that have ramifications on the issue of weaponisation are:

- **Weapons of Mass Destruction (WMD):** Article IV of the OST is the only Article among 17 that deals directly with military activities in space and, more specifically, with WMD. In paragraph 1, the contracting states "undertake not to place in orbit around the Earth any objects carrying nuclear weapons or any other kinds of WMD, install such weapons on celestial bodies, or station such weapons in outer space in any other manner." Thus, nuclear, chemical and biological weapons are prohibited from being deployed but their transit through space is not. Conventional weaponry is only banned from being placed on celestial bodies leaving its deployment in orbit around the Earth open-ended which could be subject to an individual member state's choice. As a result, Article IV has often been cited to support the claim that all military activities in space are permitted, unless specifically prohibited by the Article or any other treaty or international law.[58] The treaty, thus, does not prohibit the use of space for the entire range of passive milsats which provide ISR, communications and navigation support services to the land wars[59] evidenced by the increasing number of countries that currently operate milsats. The loosely structured Article does not prohibit testing and deploying space-based weapons or for that matter, ground-based ones aimed at satellites. Technically speaking, ASATs being non-WMD belong to the category of conventional weaponry and the legality permits their use. Given the fact that Article 51 of the UN Charter of 1945 gives a state the right to use military force to defend itself against hostile actions, is it ethics then that prevents major spacefaring powers from deploying them in space?
- **Interpretation of the Term 'Peaceful Purpose':** The core of the five principles governing outer space underscores the conduct of all human activities only for 'peaceful purposes'. Lack of definitional clarity on

such terms allows for different interpretations. The US views 'peaceful' as synonymous with 'non-aggressive' while the Soviet Union initially maintained it as 'wholly non-military'.[60] However practices over the five decades have come to interpret the term as non-aggressive, manifesting itself in the realm of almost 500 milsats designed to support Earth wars in varied roles. Joint Direct Attack Munitions (JDAMs)[61] which are guided by a GPS feed from a satellite use the space asset in a support role but are unarguably air-launched aggressive weapons by themselves. Thus, space retains its status as a sanctuary only in so far that no weapons are known to be deployed there. It becomes apparent that the OST attempts to achieve its truism in generic concepts and ideas but fails to deliver on realism.

- **Declaring of Intentions:** Article XI of the OST requires states to inform the UN, the public and the international scientific community, of space activities conducted by them as soon as feasible and practicable. States are not obliged to disclose the true function of a satellite, allowing this lacuna to be exploited to the extent that not a single registered satellite has its function listed as 'military'.[62] Given this fact, have the functions of known milsats been listed as civil, commercial or dual-use? The Registration Convention of 1974 requires each launch to be registered at the UN by the launching state. As of January 2011, 93.5 per cent of operational satellites had been registered at the UN.[63] That leaves about 70-odd unregistered and unaccounted for satellites out of a total of approximately 1,100 currently operational. It would be stating the obvious that these would be milsats, may be even of super-sensitive role and mission, which raises the question: if other milsats could be disguised as civil or commercial during registration, what led the launching state(s) to abrogate their obligation to the UN by 'denying' the existence of these satellites? Is this an exploitation of inadequately and ambiguously structured space laws or is it more likely lack of ethics on the part of the launching states?
- **Laws that Lack Teeth:** Laws that are couched in ambiguous and nuanced legalese are more likely not to be implemented effectively. Article IV contains text such as "state parties *undertake* not to place in orbit..."

which appears to be weaker in construction as against the more rigorous and stringent word "*shall*" and does not reflect an obligation binding upon the signatories.[64] The legalese of the agreements has been found to be such that it appears to offer an escape hatch to member states allowing signatories to view them from an angle that suits their perspective and emboldens them in their activities in space which are not always in-sync with what is globally understood.

Conference on Disarmament (CD), 1979

The lacunae in the existing space laws prompted some countries to put forward proposals for a more efficient space security regime. The USSR proposed a draft treaty at the UN in 1981 to ban the orbiting of any kind of weapon in outer space and on celestial bodies. This was introduced in the CD[65] in 1983, with a broader mandate that also included a ban on testing and orbiting ASATs. China proposed a ban in 1985 on all military uses of space while India and Pakistan proposed restrictions on testing and deployment of ASATs. Based on an earlier initiative in 2002, China and Russia, in 2008, jointly proposed a treaty on the Prevention of the Placement of Weapons in Outer Space (PPWT) to the CD Committee on Prevention of an Arms Race in Outer Space (PAROS) and then in 2012, reminded the CD of the draft treaty submitted in 2008. Israel has been invariably abstaining from voting on PAROS while the USA has consistently voted against such resolutions and now in May 2013, along with Canada, announced boycotting disarmament forums at the CD till June 23, 2013, while Iran presides over the CD. The USA maintains that there is no arms race in space and the existing arms control mechanisms adequately deal with non-weaponisation of space. Lack of contributions from member states of the CD and opposition by the US to almost all the proposals on PAROS are the reasons why the treaty on PAROS has remained deadlocked till date. The lack of consensus and progress on the PPWT proposed by China and Russia has resulted in the unwillingness of the two countries to support the International Code of Conduct (CoC) for Outer Space Activities proposed by the European Union in 2008, which is currently under consultations. Their unwillingness to support the CoC also stems from the fact that unlike the PPWT, the CoC does not explicitly prohibit weapons in space and one of its general principles allows for the inherent right of self-defence under the UN Charter.

An ever increasing number of countries is taking to leveraging benefits that accrue from assets in space. Countries that have recently arrived on the space-scene now look at leveraging the military-support feature of this medium in pursuance of their national security objectives. Space-based systems have enhanced the battle awareness among the terrestrial commanders and established the need for a 'full spectrum' support in real-time communications, navigation and precision guidance, reconnaissance and intelligence gathering, weather monitoring and early warning. This scramble for a slice-of-the-space-pie gives rise to a very complex and competitive geopolitical situation where security concerns of protecting their space assets may tempt countries to raise the bar from militarisation to weaponisation of space.

Space development is the norm today. The civilian role that satellites play in the service to humanity in just about every facet of modern day living far transcends the dividends that accrue in support of the military. The issue gathers added complexity as the number of stakeholders steadily rises. Not only do countries come together in their endeavours in space, participation by commercial entities and conglomerates adds a new dimension that highlights the importance of updating the 'rules of the road' for the fourth dimension. Militarisation of space is an accepted fact and is here to stay. Weaponisation, on the other hand, is not yet established but appears to be a definite trendline on the geopolitical horizon unless arrested as an immediate need.

Just as atmospheric vacuum characterises space, there appears to be a legal vacuum when it comes to weaponisation in space. The UN legislated Outer Space Treaty bans weapons of mass destruction from being tested or deployed in space but it is woefully inadequate in applying similar statutes on conventional weaponry in space. Subsequent proposals to strengthen space security and make it binding on member states to stringently abide by them have not seen the light of the day. As each conflict places a heavier reliance on force-enhancement from space, absence of legally binding instruments and norms to prevent further escalation, may well see the fourth dimension not just as that of 'supporting a war' but of becoming a battlefield itself.

Activities in space, as advocated by the international treaties and agreements, are themed on voluntary and responsible behaviour by member states. While states do so voluntarily, not much exists in real terms of mechanisms that can monitor, verify or enforce the country's obligations towards space security and sustainability. If the contemporary legal regime cannot enforce prohibition of arms in space, it would auger well for member states to exhibit diplomatic maturity in bringing in legislation that does and in the interim, from an ethical point of view, treat space as *res communis* – owned communally by mankind.

Notes

1. Bert Chapman, *Space Warfare and Defense: A Historical Encyclopedia and Research Guide* (California, USA: ABC-CLIO, Inc., 2008), Ch. 1, p. 39.
2. Joan Johnson-Freese, *Space as a Strategic Asset* (New York, USA: Columbia University Press, 2007), Ch. 1, p. 19.
3. Michael Sheehan, *The International Politics of Space* (Oxon UK: Routledge, 2007), Ch. 6, p. 98.
4. "Another Eye in the Sky", *Hindustan Times*, August 9, 2012.
5. William C. Martel and Yoshihara Toshi, "Averting a Sino-US Space Race," *The Washington Quarterly,* Autumn 2003, p. 26.
6. Chapman, n. 1, Ch. 2, p. 97.
7. Dwayne A. Day and Colin Burgess, "Monkey in a Blue Suit," *Spaceflight*, vol. 48, no. 7, July 2006, p. 266. Corona was the codename of the CIA's project, partly funded from US Air Force's Discoverer programme which was to launch photo-reconnaissance satellites over the Soviet Union.
8. David D. Chen and Molly K. Macauley, "Commercial Space Actors," in Eligah Sadeh, ed., *The Politics of Space* (London: Routledge, 2011), Ch. 5., p. 114.
9. Union of Concerned Scientists, "UCS Satellite Database," December 1, 2012, accessed at http://www ucsusa.org/nuclear_weapons_and_global_security/space_weapons/technical_issues/ucs-satellite-database.html on June 27, 2013.
10. Valery Babintsev, "The Peaceful and Military Development of Space," in Alexei Arbatov and Vladimir Dvorkin, eds, *Outer Space-Weapons, Diplomacy and Security* (Washington, DC: United Book Press, 2010), Ch. 2, p. 27.
11. Nordin Yusof, *Space Warfare: High Tech War of the Future Generation* (Malaysia: Penerbit University Teknologi Malaysia Press, 1999), p. 643.
12. Wg Cdr Kaza Lalitendra, *Militarisation of Space* (New Delhi: KW Publishers Pvt Ltd, 2010), p. 12.

13. J. R. Wilson, "Satellite Communication Key to Victory in Iraq," August 1, 2003, accessed at http://www. militaryaerospace.com/articles/print/volume-14/issue-8/ news/satellite-communication-key-to-victory-in-iraq.html on May 13, 2013.
14. Lalitendra, n. 12, p. 66.
15. Sheehan, n. 3, Ch. 7, p. 117.
16. Sadeh, ed., n. 8, p. 323.
17. Wilson, n. 13.
18. Hon. Peter B. Teets, Under Secretary of the US Air Force, "National Security Space in the Twenty-First Century," *Air and Space Power Journal,* Summer 2004, p. 1.
19. This at best is conjecture, as there is no verification mechanism to confirm or deny such an occurrence. It is unlikely that the two superpowers would invest tremendous amounts of research activity and funding, and then not have the capability secreted away piggybacking on a mother satellite in orbit, for its use in times of crisis in the future.
20. NATO, *Weapons in Space and Global Security,* Report of the NATO Parliamentary Assembly Sub-Committee on the Proliferation of Military Technology, 2003, accessed at http://www.spacesecurity.org/SpaceSecurity Report2012.pdf on May 15, 2013.
21. RIA Novosti, "Russia's Space Defences Stage a Revival," October 4, 2006, accessed at http://www.spacesecurity .org/SpaceSecurity Report2012.pdf on May 15, 2013.
22. Petr Topychkanov, "Features of the Outer Space Environment," in Alexei Arbatov and Vladimir Dvorkin, eds., *Outer Space-Weapons, Diplomacy and Security* (Washington, DC: United Book Press, 2010), Ch. 1, pp. 8-9.
23. Peter L. Hayes, "National Security Space," in Sadeh, ed., n. 8, Ch. 2., pp. 35-36.
24. Cesar Jaramillo, ed., "Space Security Index 2012," accessed at http://www. spacesecurity.org/ SpaceSecurityReport2012.pdf on May 15, 2013.
25. Ibid.
26. "Cosmos-2473 – Garpun-Type Relay Satellite," September 21, 2011, accessed at http://russianforces.org/ blog/2011/09/cosmos-2473_-_garpun-type_rela.shtml on May 20, 2013.
27. Jaramillo, n. 24, pp. 123-124.
28. Joseph A. Angelo, Jr, *Encyclopedia of Space and Astronomy* (New Delhi: Viva Books Private Limited, 2010), p. 597.
29. The International Institute for Strategic Studies, *Military Balance 2008* (London: Routledge, 2008), p. 29.
30. Sheehan, n. 3, Ch. 6, p. 107.
31. Development, Concepts and Doctrine Centre, Ministry of Defence, UK, *The UK Military Space Primer* (Wiltshire: DCDC, 2010), Ch. 3, p. 3-46.
32. Chapman, n.1, Ch. 2, p. 129.

33. "US Newest Missile Warning Satellite Encapsulated in Launch Vehicle Payload Fairing," March 15, 2013, accessed at http://www.spacewar.com/reports/US_Newest_Missile_Warning_Satellite_Encapsulated_in_Launch_Vehicle_ Payload_Fairing_999.html, May 16, 2013.
34. Pavel Podvig, "History and the Current Status of the Russian Early Warning System," *Science and Global Security*, vol. 10, 2002, pp. 21-60.
35. Pavel Podvig, "Reducing the Risk of an Accidental Launch," *Science and Global Security*, vol. 14, 2006, p. 35.
36. Yahya A. Dehqanzada and Ann M, Florini, *Secrets for Sale: How Commercial Satellite Imagery will Change the World* (Washington, DC: Carnegie Endowment for International Peace, 2000), cited in Jaramillo, ed., n. 24.
37. Jaramillo, n. 24.
38. Jonathan Amos, "Soyuz Launched Sharp-Eyed Pleiadus Satellite," *BBC*, December 16, 2011.
39. "Israel Launches New Satellite to Spy on Iran," *The Guardian*, January 21, 2008.
40. James Clay Moltz, *Asia's Space Race* (New York: Columbia University Press, 2012), pp. 10-11.
41. Jaramillo, n. 24, p. 124.
42. ISRO, "Cartosat 2B," accessed at http://www.isro.org/satellites/cartosat-2b.aspx on October 15, 2012.
43. http://www.bbc.co.uk/news/technology-16337648 accessed on May 2, 2012.
44. http://www.nasaspaceflight.com/2012/04/china-launch-compass-duo-long-march-3b/, accessed on May 2, 2012.
45. Chuang Shi, QiLe Zhao, Min Li, WeiMing Tang, ZhiGang Hu, YiDong Lou, HongPing Zhang, Xiao Ji Niu, Jing Nan Liu, "Precise Orbit Determination of Beidou Satellites with Precise Positioning," *Science China Earth Sciences*, vol. 55, issue 7, July 2012, pp. 1079-1086.
46. "Galileo Navigational System Enters Testing Stage," October 13, 2012, accessed at http://www.dw.de/galileo-navigational-system-enters-testing-stage/a-16304096-1
47. Peter B. de Selding, "European Commission Urges China Dialogue," April 5, 2011, accessed at http://www. spacenews.com/article/european-commission-urges-china-dialogue#.UaMCtPi6YdU on May 27, 2013.
48. A. Bhaskaranarayana, ISRO, "Indian IRNSS and GAGAN," presentation to COSPAR meeting at Montreal, July 15, 2008, accessed at http://www.oosa.unvienna.org/pdf/icg/2008/expert/2-3.pdf on October 29, 2012.
49. ISRO, "Future Programmes," accessed at http://www.isro.org/scripts/futureprogramme.aspx on October 29, 2012.
50. ISRO, "IRNSS-1A," accessed at http://www.sac.gov.in/SACSITE/IRNSS-1A.html on May 21, 2013.

51. Stephen Clark, "Japan to Build Fleet of Navigation Satellites," April 4, 2013, accessed at http://www. spaceflightnow.com/news/n1304/04qzss/#.UZtQy_i6YdU on May 21, 2013.
52. http://www.jaxa.jp/projects/sat/qzss/index_e.html, accessed on April 10, 2012.
53. Lalitendra, n. 12, p. 133.
54. John H. Cushman, Jr., "US Satellite Plans Falter, Imperiling Data on Storms," *The New York Times*, October 26, 2012.
55. Turner Brinton, "Outlines of Pentagon Weather Satellite Plan Starting To Emerge," June 30, 2010, accessed at http://www.spacenews.com/article/outlines-pentagon-weather-satellite-plan-starting-emerge#.UZyeK_i6YdU on May 22, 2013.
56. Marc V. Schanz, "Weather Satellite Reboot," January 12, 2012, *The (US) Air Force Magazine*, accessed at http://www.airforcemag.com/DRArchive/Pages/2012/January%202012/January%2012%202012/ WeatherSatelliteReboot.aspx on May 23, 2013.
57. Keith Hayward, "Europe in Space," *The RUSI Journal*, vol. 155, no. 6, 2010, p. 39.
58. Ranjana Kaul, "Leveraging Outer Space for National Security," *Air Power Journal*, vol. 2, no. 4, Winter 2007, October-December, p. 130.
59. Defence Intelligence Agency, *Soviet Military Space Doctrine* (Washington, DC: DIA, 1984), p. 11.
60. Kaul, n. 58, p. 129.
61. JDAM is an guidance kit that is strapped onto an unguided air-delivered bomb which then guides the 'smart' bomb to its target using its integrated inertial guidance system coupled to a GPS receiver.
62. Kaul, n. 58, p. 133.
63. UN OOSA, "Registration of Objects Launched into Outer Space," accessed at http://www.oosa.unvienna.org/ oosa/SORegister/regist.html on June 6, 2013.
64. Dr. Harnam Bhayana, *International Law in The Regime of Outer Space* (Calcutta: R. Cambray & Co. Pvt. Ltd., 2001), Ch. 6, p. 267.
65. The CD is a forum established by the multilateral international community, including non-government organisations to address issues on arms control and disarmament. Though not formally a part of the UN, a representative of the UN Secretary General serves as the Secretary General of the conference and activities of the CD are reported annually to the UN.

8

Some Geostrategic Imperatives for India

Even if you're on the right track, you'll get run over if you just sit there.

— Will Rogers
American Humorist, 1879-1935

Ranking sixth in the world in terms of space budget and technological capabilities, with over US$ 26 billion invested in space assets, India today ranks amongst globally established space players. Home to the largest civil constellation of remote sensing satellites in the world, India banks on its robust and reliable space workhorse, the PSLV, to loft 1.6 ton satellites into LEO and about one ton into GEO. The PSLV is a proud symbol of the country's self-reliance and a measure of technological maturity by way of achieving 23 consecutive successful launches out of 24 to date. This space workhorse has not only showcased the indigenous launch capability for meeting requirements of the domestic space programme but also in the global commercial field by placing 35 foreign satellites into orbit.

The nation has undertaken varied activities in space besides orbiting satellites in LEO and GEO. Its mission to the Moon in 2008 made the world aware of existence of water on it. Mangalyaan, the mission to Mars launched in November 2013 in currently enroute to Mars and scheduled to enter Mars' orbit in September 2014. The Space Capsule Recovery Experiment in 2007 can be deemed to be the precursor to manned space flights in the times to come. Satellite-based navigation services are being fielded in projects like the IRNSS and GAGAN. The array of such diverse activities bears testimony to the nation's multi-dimensional space programme evolving over the years and standing today self reliant and technologically robust.

This chapter offers an insight into some of the challenges likely to be faced by India in its endeavours in space, at both the domestic and the global geopolitical and geostrategic levels. It is the latter field that offers tremendous scope in converting technological achievements into geopolitical spin offs. This concluding chapter tries to identify the drivers that are likely to shape the evolving global order in the years to come and offers some viable options and alternatives in the Indian context to leverage them into national capability building to suit its national interests.

Bonafide Need for India's National Space Policy

Despite there being no national space policy till date, the indigenous space programme has had a focussed direction since its inception. Commencing its odyssey in space with the pivotal aim of socio-economic development, the role of space has traditionally followed this course. The Space Commission, operating directly under the Prime Minister, formulates and oversees policies at the apex level while the Department of Space implements the programmes through ISRO and other subsidiaries.

The Indian space programme is guided by the Allocation of Business Rules for the Department of Space, along with related legislations and regulations of the government and individual policies on remote-sensing data, satellite communications, mapping, etc.[1] What remains lacking is a comprehensive national space legislation or a specific law that deals with space activities undertaken by the country in the contemporary context, which could also factor in activities in the years to come. The reason for not legislating such a domestic law could well be the fact that the Constitution was adopted in 1950, well before forays into space which commenced in the 1960s. The country is considered to be slow in ratifying or acceding to treaties and takes its own time to incorporate its convention obligations into domestic law.[2] Some are also of the view that the space activities till 2000 were purely in the government's domain with no private partnership and, thus, did not necessitate articulation of a domestic space law.[3]

Issues that justify the need for a national space policy have been covered in detail in Chapter 6. Two of the reasons merit being reiterated briefly to further the advocacy of such policy. The first is the ever increasing participation of the country's private sector in ventures in space. Commencing early 2000, the domestic space sector has been opened up for participation by commercial private enterprises. Their role, as guided by Antrix, the commercial arm of ISRO, grew significantly with over 500 companies participating in the space effort,[4] with almost 60 per cent of launch vehicle cost flowing back to the industry.[5] Launch services apart, proliferation of civilian commercial applications in the space sector necessitates a larger involvement of the private players. With the entry of private entities into satellite systems, including licensing, it is the Government of India that bears international responsibility and liability. This, in turn, dictates the need to adopt at the national level, an unambiguous, transparent and user-friendly legal regime supported by readily available information for the private enterprise—a legislation that not only regulates and governs but also has the force of law over all the stakeholders involved in the national space sector.

The second factor which adds legitimacy to the need for a national law is that of the Indian space launch sector's forays in the field of global commercialisation. Due to its relative low cost and robust reliability, the PSLV remains a very viable option for launching satellites into LEO, having orbited 35 foreign satellites so far. For all such foreign customers worldwide, India bears responsibility and liability as the 'launching state' as enshrined in UNCOPUOS's Rescue Agreement, the Liability Convention, and the Registration Convention. While India, having ratified all the three treaties, bears responsibility at the UN level, it would serve the country well to include in its national space law certain standardised operating protocols that identify responsibility that ought to be vested with its foreign partners, especially with respect to dispute resolution, debris mitigation, insurance and financial liability.

Soft Power Projection

As the global geopolitical milieu unfolds and changes in the years to come, the domain of space will witness ever increasing launches and employment of platforms that support a nation's collective will to further its socio-economic development. While the select-few major space players will continue to invest in space systems that contribute to their national security, the balance of the world will increasingly view space as a medium that charts their course to progress. Civil oriented space programmes will aim at bilateral and multilateral mechanisms towards collaborative ventures by sharing costs and space technologies. Such ventures will enable newer players to exploit benefits from space-based capabilities and for the established players, will enhance transparency and confidence among international partners. Large conglomerates and private entities will ever increasingly jostle for the civil and commercial space segment while government bodies are likely to focus on national concerns. This factor is likely to reshape alliances that could impact the global world order.

In the Indian context, there is much credit in exploiting the country's growing economic might and further enlarging geopolitical gains and spin-offs by engaging with foreign partners. Indian presence in the global field of partnerships and collaborations, though modest, has been on an upward trajectory since the turn of the millennium. Its share of the US$ 200 billion annual global space commerce, though meagre, has the potential of becoming significant as its technological prowess in the space domain covers almost the entire spectrum of capabilities from ground support infrastructure, launch segment, satellite manufacturing and satellite imagery, to other space-derived services. Commercial tie-ups not only enhance transnational cooperation but can also serve the country as tools for international diplomacy.

Increasing the Launch-Generation Capacity: Operating off two launch pads at the sole space complex at the Satish Dhawan Space Centre at Sriharikota Range (SHAR), the country's launch generation at best is 2–3 launches a year. This is woefully inadequate to meet the burgeoning domestic needs as well as provide launch services commercially. A project to add one more rocket assembly line at the second launch pad at SHAR awaits the government's sanction though it will not substantially increase the launch generation rate. The country can ill afford to stagnate at this launch rate and will need to look at further options to increase the same. Plans are afoot on the feasibility of a third launch pad at SHAR which will cater for the GSLV IIIs and the reusable space launcher, the *Avatar,* in the future. Yet another option could well be in generating a geostrategic option of building redundancy by establishing yet another space port and have the planned third launch pad (of SHAR) operate from there. A large number of low-latitude stretches on the east coast could yield viable venues for additional seaward launch sites. Additional launch sites would, of course, need to be complemented with the complete range of additional support facilities that make up a space launch port. In terms of geopolitical spin-offs, the country could enter collaborative ventures with countries in the subcontinent and establish launch sites at near-Equator regions. Malaysia and Indonesia both have adequate land that is sea-facing to the east and have fairly well developed domestic space assets but lack indigenous launch capability. Not only can India lease launch sites or buy rights to launch from these countries akin to arrangements that France has with French Guyana for launching its Ariane launchers from Kourou, it can build upon its soft power expansion options in assisting such countries in constructing launch and launch support infrastructure. The earliest that India can hone and operationalise its indigenous cryogenic technology for the GSLV launcher, additional options will emerge not only in orbiting heavier payloads in GEO for domestic needs, but also for launches of its foreign partners, providing additional fillip to the country's stature.

Export of Space Derived Services and Applications: The country's rather well developed space technologies in the fields of launch, communications and remote sensing can reap tremendous benefits not just in the matter of commerce but more significantly towards generating and expanding its soft power. This should specifically target underdeveloped and not-yet-developed countries in the subcontinent, Africa and Latin America by leasing transponders on communications satellites and exporting imagery. For countries with emerging economies, India should build upon its already well established competency in the launch sector and proffer the complete commercial launch capability for launching foreign satellites in LEO and GEO. The domestic capability in satellite construction of such countries could be enhanced by collaborative mechanisms that India can offer, especially in civil and commercial satellites. Such countries should be proffered the entire range of services that Antrix, the marketing arm of ISRO, currently engages in, with a view to project and expand the country's 'soft power'. Significant inroads can still be made in establishing ground infrastructure for its foreign partners. With the satellite-based regional navigation system, the IRNSS, becoming operational in 2015, another avenue for export of services opens up though limited to countries that lie in the system's subcontinental footprint.

GAGAN and its Global Utility: GAGAN, the GPS Aided Geo-Augmented Navigation system, is a satellite-based regional navigation service that is being fielded by the Airports Authority of India (AAI) and ISRO that aims to provide air traffic management and navigation over the Indian air space, making India the fourth in the world to do so. To be operational in 2014, GAGAN with its enhanced accuracy will not only make the skies and landings safer by providing a three-dimensional approach operation but also enable lesser safety-spacing between two aircraft allowing for three times the number of aircraft to fly. Better air space management is estimated to cut airline fuel costs by 20 per cent[6] providing for cheaper air travel. As GAGAN's footprint extends from the Middle East Asia to Australia and GAGAN being interoperable with the GPS, the augmentation can be provided over this region too. Going beyond what the system can offer to countries of the subcontinent within the coverage, GAGAN, when fielded, will be in a position to bridge the gap between the navigation systems of Europe (served by EGNOS) and Japan (served by MSAS) and provide continuous and seamless navigation services to long distance commercial flights. This venture should be seen not just from the commercial payoffs that will accrue but more importantly as a move that can provide a global outlook to India's national space programme and position it as a responsible and credible space player in the global space community.

Fig 8.1: Satellite-Based Augmentation Services Worldwide

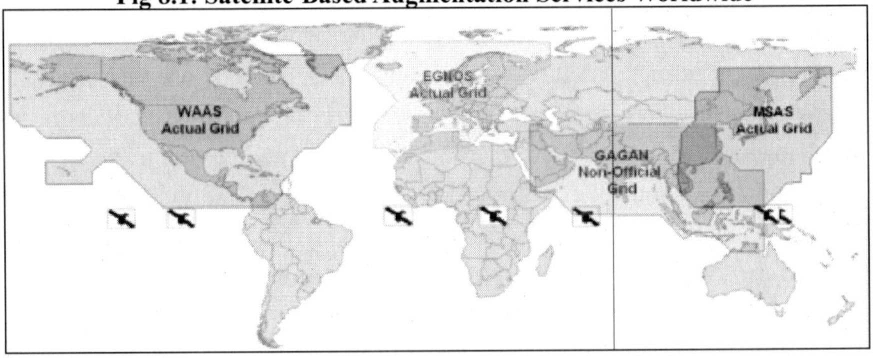

Adapted from http://www.elisanet.fi/master.navigator/InfoEGNOS_WAAS.htm

Endeavours at the International Level

Global Space Legislation

India is a signatory to most of the instruments that govern responsible conduct of space activities as legislated by the UN. Negotiations that have taken place with respect to the Outer Space Treaty and consultations at the International Code of Conduct have largely seen the country take a reactive position to proposals put forth by others. For the country to be taken seriously in the global space community, India needs to redefine its focus and participate as an engaged partner that suggests and supports legislation that is comprehensive and implementable, while projecting intention, confidence and transparency in its space endeavours. India should take the initiative and suggest a concrete action plan that delves into verification mechanisms and penalty obligations which then are implemented through the auspices of the UN. Naysayers of the ICoC argue that every country has a sovereign right to safeguard its assets in space. It would suit India's interests to suggest, without diluting the essentials of the ICoC, an 'interim' proviso that ASATs be limited to only under 250 km altitude so that the resultant debris then falls back to the Earth in a reasonable timeframe.

India appreciates the fact that the major space players have a responsibility in shaping the space security regime and no worthwhile result can be achieved unless these players demonstrate their willingness to set an example by walking-the-talk. India, while continuing to support the UN legislation, the ICoC and the Conference on Disarmament (CD), and to champion the cause of a weapons-free space, could well take the lead in establishing bilateral and multilateral mechanisms with the big three—the US, Russia and China, that have demonstrated their capabilities in ASATs—and with other significant players like the ESA and Japan. India can offer itself as an initiator and a mediator to garner international consensus for further strengthening treaties which aim to keep space a conflict-free zone.

Space Situation Awareness (SSA)

Satellite launches worldwide have created a situation that the orbital bands are increasingly getting cluttered, especially the LEO. The risk of collision with another satellite, debris or a cosmic near-Earth object rises consequentially and threatens space security. There is currently no global space surveillance system effort, largely due to the sensitive nature of the surveillance data. The US Strategic Command's Joint Space Operations Centre (JSPOC) detects, tracks, and identifies space objects through an elaborate network of 29 ground-based radars and optical sensors deployed around the world and extends this capability—at no cost to the international community—by warning countries and commercial space operators when their satellites are at risk from large space debris or other satellites. It would be in the interest of every stakeholder in the space community, including commercial, to overcome the barriers to data-sharing and promote sharing of satellite positional data and data detected from their own national space surveillance networks and feed the same into a global SSA data centre. Russia, France, Germany, China, Japan and Canada are some other countries that possess such national capability though not at the scale of the US. India should enter a collaborative mechanism wherein ISRO Telemetry, Tracking and Command Network (ISTRAC), which keeps track of satellites in LEO and the Indian Deep Space Network (IDSN), both at Bangalore, link up with the SSA programmes of other major countries in an architecture that allows sharing of space surveillance data through common communications and data exchange protocols. Besides contributing to space security and sustainability, the endeavour will help exemplify transparency in the intentions and nature of the Indian space programme and add value to confidence building measures.

International Cooperative Involvement

After hosting the 39th Scientific Assembly of the Committee on Space Research (COSPAR), with the theme "Space – For the Benefit of Mankind" which was attended by over 2,500 scientists from 75 countries in Mysore in July 2012, India is now in a position to take the cooperative and collaborative approach further. Cognisant of the fact that space ventures come with hefty price tags and are technologically challenging, a 'go alone' approach would be at the cost of some other developmental area critical to a developing country and it would, thus, be beneficial to partner with other developed and developing spacefaring countries, specifically in the field of emerging technologies, space sciences and beyond-Earth exploratory missions. The country has already explored this approach by collaborating with the Soviet Union in 1984 and having its first astronaut in space aboard the Salyut space station. In beyond-Earth forays, the Chandrayaan 1's mission to the Moon in 2008 carried the most international payloads ever in cooperation with the USA, Bulgaria and 17 countries within ESA. There lies tremendous potential in taking these initial steps further to collaborate with international partners in civil and commercial fields that are especially technologically challenging. The strategic payoffs that such ventures could harness would turn exponential in the years to come.

Space Security Imperatives for India

> *India's Space Program is liable to be ignored by strategic commentators until it visibly has a military dimension.*
>
> — Neville Brown[7]

For far too long, has India suffered the 'Pakistan-centric' myopia in its defence preparedness and yet, was caught unprepared by the Kargil intrusion in 1999. This was at a time when it had a fairly well established and proven capability in its remote sensing satellites and yet the conflict did not witness a single satellite being manoeuvred to support the ground and air battle.[8] Since Pakistan's space capability is relatively insignificant, India did not focus on security-enhancing capabilities from space or on protecting its space assets. It is only in the past decade or so that she has conceded to realign its politico-military focus towards China alarmed, by the latter's entry into the economic and military global arenas. China's ASAT in 2007 has further compounded the strategic imbalance, forcing India to revisit its geostrategic compulsions, especially in the field of its space capabilities.

National Space Security Policy: Its Time has Come

Much like the void that the absence of a national space policy creates in the global consciousness, the airing of national security imperatives that can be addressed by the domestic space programme, too, is prominent by its absence. A stated policy apart, even leveraging of the civil-oriented space infrastructure into the national security apparatus remains absent. Space assets inarguably offer tremendous opportunities and options for enabling and supporting the military capabilities of the nation. A stated policy in the form of a White Paper or a National Space Security Policy (NSSP) which ideally lays out the security environment faced by the nation and then lists out the security imperatives that could be addressed by space capabilities wholly or in part, as yet, remains to be documented. Having identified the need of a national space policy above, it would be a corollary that a policy on the security dimension from space should naturally evolve from it. The NSSP should aim to protect national security interests critical to space security. Besides exhibiting good intentions and transparency to the world, such a policy should, at the domestic level, spell out the role and charter of each and every stakeholder, crystallise institutional processes and clear the ambivalence on the interplay between various participating agencies, including the civil and commercial.

National Security and the Societal Percept

To a very large extent, a policy on the national space security would be able to allay apprehensions in the minds of the domestic populace, especially the academia. Almost all spacefaring nations developed ballistic missiles and then had their space launchers evolving out of their missile programmes. In fact, even within their space programmes, the military-use of space came about before the civil-use. Thus, their societies find military ambitions in their space programmes quite acceptable and a part of the evolution of space exploits. The Indian societal percept, on the other hand, finds succour in the long-held Gandhian idealism of peaceful exploitation of technologies that aim at nation building. Adding a military hue to the motif of non-violence then, is frowned upon as rhetorical sabre-rattling. A well articulated space security policy can be the driver in changing the social percept – a policy that bridges the so-far held position of 'idealism' with the pragmatic need today of 'realism' - enabling the social milieu in better understanding security dimensions of the contemporary geostrategic realities faced by the nation, parts of which could be addressed from space, *without needing to weaponise it.*

The Promise of a Joint Space Command

The communications satellite GSAT 7 was launched into GEO on August 30, 2013 aboard the Arianne. While this was the 71st Indian satellite to be orbited, its uniqueness lies in the fact that the GSAT 7 is the first to be publicly championed as a pure military satellite – one which is dedicated for the communications needs of the Navy. The launch heralds an era of leveraging space assets for national security as this will be followed by two more dedicated satellites, one each for the Air Force and the Army. As of date, the intelligence agencies of the three Services, through their respective space cells in the Service Headquarters, receive processed satellite data from the tri-service Defence Image Processing and Analysis Centre (DIPAC) in New Delhi.[9] There has been much conceptual loud-thinking on a tri-Service organisation which needs to be fielded to undertake and coordinate military aspects of the space infrastructure. An integrated Joint Space Command (JSC) is the organisational platform which needs to be established – an organisation that should be manned and run by members of all the three Services in a synergised manner. This would help to leverage operational payoffs in the national, strategic and tactical domains holistically, duly signifying space as the fourth dimension and not a domain specific to a single Service. The JSC should be placed under the control of the apex Integrated Defence Staff (IDS)—the headquarters of which also have the three Services under it – through the recently established coordinating agency, the Integrated Space Cell. As and when additional capabilities are fielded through satellites planned for the future, the JSC then will be better placed to expand its role in defence communications, battle management, early warning, navigation, reconnaissance, border surveillance, targeting and intelligence including electronic. Since the formation of the JSC will be the first step towards capability building towards security issues, the lead-Service should ideally be rotational to begin with. Once up-and-running, the call to identify a Service to permanently steer it could then be taken, if deemed necessary.

Energising Space Capabilities for Strategic Security

Amongst the currently orbiting satellites are some which could be classified as dual-use ones. Operating in LEO, the civil satellites Cartosats 2A and 2B provide optical images of resolution of up to a metre. This image intelligence capability is buttressed with SAR capability which sees through clouds and fog in day and night operations, a capability demonstrated in RISATs 1 and 2. Of special utility is the strong echo returned from metallic objects, denying the adversaries' battle-platforms the security and protection of camouflage. For the role of force-enhancement, a judicious mix of dedicated military satellites for optical-imaging, multi-spectral, ELINT, SIGINT and SAR will provide applications for ISR that cover the entire spectrum from strategic to tactical support to the military. A cost-effective option could even consider a *combo* payload which carries SAR and spectral-imaging sensors on the same satellite. Even if this surveillance capability is kept restricted to the subcontinental domain, the nation needs to have in orbit a larger fleet of manoeuvrable LEO satellites which enables shorter revisit times over a specific area. Of special strategic advantage will be to establish a 'persistent look' capability – a capability that provides continuous surveillance over a designated area of interest for a specified period of time.

In the domain of communications from GEO, GSAT 7 is the first dedicated military satellite for communications while military communications have so far been supported by the inherently civil space programme, albeit in a limited manner by way of a few transponders aboard some of the civil communications satellites. A dedicated satellite for each of the three Services is operationally inescapable to overcome challenges of network-centric warfare in the contemporary military environment. The rising demand for bandwidth will additionally necessitate earmarking dedicated transponders on selected civil satellites for the military. The GEO is especially beneficial to perform the function of launch detection of ballistic missiles. India has stated its plans to orbit such an early-warning radar satellite to maximise the reaction time between hostile launch detection and intercept sequence.[10] The IRNSS when in operation should be capitalised on to provide indigenous satellite-based navigation aid for precision targeting systems for the three Services.

Two factors will have a big role to play in bringing the above wish-list to fruition. The first is the need to increase indigenous launch-generation capacity as a primacy, which requires almost immediate addressing. The second factor that will require prioritisation is satellite manufacture and launch. This aspect will need prudent handling of competing demands for space assets between those in support of the military and those in support of socio-economic nation building. Having honed the technologies necessary for a fairly robust civil space programme, space capabilities required dedicatedly for the military will only be an extension of those already achieved and should not pose any technological challenges.

Another significant force-enhancement feature of India's national space security paradigm that should be leveraged is the 'reactive-launch' capability which accords a threat-based capability to launch-on-demand. ISRO has already demonstrated this capability, albeit limited, by launching RISAT 2 in the aftermath of the Mumbai blasts. ISR support to the military can gain a significant fillip by launching lightweight satellites at short notice by quick-to-assemble-and-launch mobile launchers. This accords an added advantage, that of increasing domestic launch-generation capacities.

Indian ASAT – Going Beyond Rhetoric

> *Unless India stands up to the world, no one will respect us. In this world, fear has no place. Only strength respects strength.*
>
> — A.P.J. Abdul Kalam

A very significant capability that needs to be factored in for security of space assets is that of the oft-debated ASATs. The ASAT option and that too of a debris-creating kinetic-kill or direct-impact variety is a serious space security dilemma that India faces. Should India, in pursuance of the Gandhian ideology of abstaining from aggressive employment of a technology that so far has reaped tremendous benefits to its society, continue to maintain the status of a responsible space player? Or, address the vulnerabilities of its space infrastructure by developing an ASAT capability which then creates a strategic balance in the Indian subcontinent?

The Need for ASAT Capability: Indian space assets are vulnerable to ASAT weapon systems because the neighbourhood possesses one.[11] China's testing of an ASAT in 2007 has unfortunately upped the ante and tilted the rhetoric in favour of India wanting to balance the asymmetry with the better space-enabled regional rival. An Indian ASAT programme is an option to establish deterrence in its geopolitical equation with China. Such a capability, duly advertised and demonstrated, should reduce adversarial attempts at degrading India's exploitation of the medium of space. The capability will also position the nation at an exclusive place in the global space order by acquiring respectability among members of the spacefaring club. From this position of strength, India then should prevail upon the naysayers into arriving at a consensus at international agreements on the CD and PAROS. While major space powers might extol the virtues of application of military force 'in, from and through space', the Indian narrative at best has to be limited to military force enhancement, to include a defence oriented ASAT capability that protects its space assets. Towards the same, V. K. Saraswat, Director General of the Defence Research and Development Organisation (DRDO), stated in 2011, "We have all the technology elements which are required to integrate a system...(to) defend our satellites" through the country's missile defence programme.[12] A reality check on the efficacy of a Ballistic Missile Defence (BMD) operating as an ASAT weapon would suggest that currently there exist gaps in the technologies required to bridge the capabilities of the two. The last BMD test conducted in November 2012 involved an interceptor with a speed of 4.5 Mach destroying a missile-target at an altitude of 15 km while the next phase of testing will involve an interceptor with a speed of 6-7 Mach.[13] The highest altitude of interception achieved among seven tests so far has been 80 km. To take this capability into the realm of space deterrence in LEO will require a major technological leap, not just in terms of altitude of interception but, equally importantly, in the speed of the 'target' being intercepted. Taking a leaf from the narrative of the US' ASAT programme, which had ASAT technology as an ancillary capability to its BMD programme,[14] the Indian BMD programme too can be leveraged as a technology demonstrator for an ASAT capability.

The Cons to ASAT Testing: The reasons advocated by pacifists are as numerous and significant as those being propounded by the hawks. Some of the issues which contribute to the decision dilemma and require factoring in are:

- Worldwide condemnation as an aftermath of the test is a certainty. China's ASAT in 2007 witnessed global consternation, more so due to the additional debris that was created from the kinetic interception at 865 km. The US followed a year later with yet another ASAT test, getting away with the justification that the altitude of interception, at 250 km, was low enough to cause the debris to decay and fall back to the Earth. But for the fact of the Chinese debris, both tests were within the permissible activities enshrined in the UN's relevant space legislation.
- India will lose its moral high ground. The traditional stance of advocating non-weaponisation of space will vitiate its political importance and reduce elbow room in any negotiations on the anti-weaponisation platform in the future.
- The image of a responsible spacefaring nation that India projects will take a hit. The resultant undermining of its credibility will reduce its leverage in the domain of generating soft power.
- It will inflame regional tensions and encourage others to follow suit.[15] Pakistan for one is likely to match this capability as has been the historically proven norm. Notwithstanding the fact that Pakistan has no space launch capability, its does indeed have ballistic missiles and from its all weather friend China, a more-than-covert support of an established space power.
- Upping the ante in the sphere of building technologies for an ASAT weapon will certainly be at the cost of its civil oriented space ventures with a concomitant impact on economic development. The factor of desirability versus affordability will retain its relevance in a developing economy.

Options for ASAT Testing: In any case, the decision for testing an ASAT system should be guided by an articulated space security policy based on the contemporary geostrategic realities and should not rest with, or be driven by, space technologists. A political positioning for defence of national space assets will then throw up options for testing of an ASAT weapon system essentially in identifying the ideal altitude of interception such as:

- *Interception in LEO*: Most of the Indian satellites are launched into LEO, orbit in the altitudes of 600-900 km. The IRS 1 series (-P3, -P4, -1B, -1C and -1D) have lived their operational life and could be used as a target for the ASAT weapon. It is this altitude band that creates long lasting debris and poses a potential threat to satellites worldwide. China's ASAT test in 2007 at 865 km altitude created 150,000 pieces of debris of the size of one centimetre or more which is estimated to remain in LEO for 100 years.
- *Low-LEO Interception:* Interception occurring at altitudes up to 300 km is relatively safer from the point of view of life of the debris. The ASAT test conducted by the US in 2008 destroyed a defective spy satellite at an altitude of 250 km whose debris decayed and fell back to the Earth within 40 days. India currently operates no satellite at such low altitudes – altitudes which ideally support operations of military satellites in the ISR role.
- *Dummy Target in Low-LEO:* To maximise the advantages of an ASAT test at altitudes below 300 km, a dummy payload can be launched aboard the PSLV and placed in orbit which then acts as the target for the ground-based ASAT interceptor. This option remains the most pragmatic and India can take this a step further by notifying the concerned UN body of such an event prior to the conduct. This would be in keeping with the provisions of Article 9 of the Outer Space Treaty which stipulates, "States must conduct international consultations before proceeding with activities that would cause potentially harmful interference with activities of other parties..."

Other Prerequisites

All the options for demonstrating an ASAT weapon system pre-suppose the existence of more than adequate technological capability in surveillance and tracking of objects in space. Regardless of the option chosen, the infrastructure required for accurate fixing of target-satellite will require an extensive network of ground-based tracking stations which ideally should be supplemented by surveillance-satellites in orbit. The ISRO Telemetry, Tracking and Command Network (ISTRAC) at Bangalore provides telemetry control and keeps track of Indian satellites in LEO while the INSAT Master Control at Hassan does the same for satellites in GEO. This infrastructure will need to be upgraded and enhanced to enable surveillance and cataloguing of the space assets of the adversaries. Such a capability would aim to analyse satellite or threat characteristics, capabilities, intent and current disposition, to eventually arrive at timely and actionable intelligence to negate the threat.

The spectrum of ASAT capabilities need not be centred around a kinetic-kill interceptor alone. Light weight co-orbiting satellites is yet another option though these would be classified as a weapon-in-space. Other ground-based hard-kill weapons which should be explored are directed energy weapons and lasers. Capabilities of ground-based soft-kill options too should be considered for development, which should include electronic attacks on satellites by jamming and spoofing. Such capability enhancements should be planned to be developed and fielded in an incremental manner.

Regardless of the decision to demonstrate ASAT capability, India still needs to implement mechanisms to protect and harden its space assets. Electromagnetic and radiation hardening of satellites and its components, anti-jamming capabilities and orbital manoeuvring are some of the measures that should be implemented on priority, for at least some critical high-value satellites.

India is among a handful of nations in the world possessing a robust and credible multi-dimensional space capability. Its space programme is based on duality of principles – the first that aims for development on the national socio-economic front and the second, that of autonomy. The motives of its space programme have undergone a paradigm shift by way of broadening into a vision – a vision that appears to have naturally gravitated from that of human and societal development to a stance that views India's presence in space as a measure of global prestige. The robustness of the space workhorse, the PSLV, has enabled the nation to foray into the commercial launch sector while international collaborative ventures in satellite construction broaden the spectrum of bilateral tie-ups. Commerce is not necessarily the only motivator as such partnerships with global players offer tremendous foreign policy spinoffs in India's favour and serve the country as tools for international diplomacy.

Major space-launch capable nations would aim to deny access to space by the others. Technology-denial will aim to garner for themselves larger pieces of the space commerce pie, including the potential of harnessing space solar energy and resources from the Moon in the future. Where will India position itself in the ever changing global space order? How will it strategise to maintain its status in the decades to come? What shape will its national space policy take to address current realities and likely trendlines of the future?

Do India's geostrategic security concerns merit inclusion in its space matrix? Intrinsic in its desire to project itself as a peaceful nation, India should temper its ideological stance and allow prudence to suggest pursuing its national security goals, as the regional and continental realities provide adequate rationale to enfold its space security concerns into its space programme. Given the geopolitical environment in its neighbourhood, use of space for national defence finds legitimacy.

Militarisation of space is here to stay. Is weaponisation of space the next logical outcome? No nation has as yet placed a weapon in space. While continuing to pursue the issue of weapons-free space in the international fora, would India be compromising this stance by going ahead with an ASAT test to showcase the technological capability of protecting its space assets that results in space deterrence and regional balance? Or, to balance both the conflicting requirements, does it need to engage in political manoeuvring to manage international opinion in its favour?

Answers to most of these questions will crystallise once the bigger question of India's regional/ continental/global ambition is answered. Where does India want to position itself economically and militarily in the decades to come?

Notes

1. Minister of State's reply in Lok Sabha on February 23, 2013, to Unstarred Question No. 369 on Indian activities in space, accessed at http://www.isro.org/parliament/2011/Budget/LUSQ369.pdf on March 6, 2013.

2. "India and Space Laws: A Millennium Perspective," March 9, 2000, *The Hindu*, accessed athttp://www.hindu. com/2000/03/09/stories/08090008.htm on March 7, 2013.
3. Dr. Ranjana Kaul, in a presentation titled "National Space Legislation: A blueprint for India," at the 3rd Regional Space Conference on Bringing Space Benefits to the Asia-Pacific Region, Bangalore, June 27-29, 2005.
4. Brian Harvey, Henk H. F. Smid and Theo Pirard, *Emerging Space Powers- The New Space Programs of Asia, the Middle East, and South America* (Chichester, UK: Praxis Publishing Ltd., 2010), p. 250.
5. "*Overview of Indian Space Sector 2010*," accessed at http://mycii.in/KmResourceApplication/E000000056. 662.Antrix-CII-Deloitte%20report%20on%20Indian%20Space%20Sector%202010.25Aug10.pdf on October 19, 2012.
6. "Air Navigation System GAGAN to Help Jet Airways, Indian Airlines, Others to cut Fuel Cost by upto 20%," *The Economic Times*, October 8, 2012, accessed at http://articles.economictimes.indiatimes.com/2012-10-08/news/34323085_1_indian-skies-air-navigation-airlines-end on October 26, 2012.
7. Neville Brown, *New Strategy Through Space* (London: Leicester University Press, 1990), p. 28.
8. Sqn Ldr K.K. Nair, *Space: The Frontiers of Modern Defence* (New Delhi: Knowledge World, 2006), p. 17.
9. Cmde Ranjit Rai, "Indian Navy: A C4ISR Nuclear Force," *India Strategic*, August 2010 accessed at http://www.indiastrategic.in/topstories717.htm on August 30, 2013.
10. "India Adding Space Radar to Missile Shield by 2014," *Space News*, March 16, 2009, p. 9.
11. Air Chief Mshl P. V. Naik, quoted in Victoria Samson, "India's Missile Defence/Anti-satellite Nexus," *Space Review,* May 10, 2010.
12. Indrani Bagchi, "India Working on Tech to Defend Satellites," *The Times of India,* March 6, 2011.
13. Rajat Pandit, "India Tests Missile Shield, DRDO Says it will be Operational by 2014," *The Times of India*, November 24, 2012.
14. Keith Stein, Secure World Foundation, "Group Believes India will Perform Anti-Satellite Test," March 8, 2011, accessed at http://www.examiner.com/article/group-believes-india-will-perform-anti-satellite-test?CID=examiner_alerts_article on September 17, 2013.
15. Editorial, "India's Anti-satellite Comments Illustrate Underlying Reality," *Space News*, January 18, 2010, p. 18.

Index